A First Course in Topological Field Theory

UNIVERSITY LECTURE SERIES VOLUME 80

A First Course in Topological Field Theory

Jürgen Fuchs
Christoph Schweigert
Lukas Woike

AMERICAN MATHEMATICAL SOCIETY
Providence, Rhode Island

2020 *Mathematics Subject Classification.* Primary 81T45, 18M20, 57R56.

For additional information and updates on this book, visit
www.ams.org/bookpages/ulect-80

Library of Congress Cataloging-in-Publication Data

Names: Fuchs, Jürgen, 1957 May 19- author | Schweigert, Christoph author | Woike, Lukas, 1991- author

Title: A first course in topological field theory / Jürgen Fuchs, Christoph Schweigert, Lukas Woike.

Description: Providence, Rhode Island : American Mathematical Society, [2026] | Series: University lecture series, 1047-3998 ; volume 80 | Includes bibliographical references and index.

Identifiers: LCCN 2026002056 | ISBN 9781470485078 paperback | ISBN 9781470485504 ebook

Subjects: LCSH: Quantum field theory | Differential topology | Manifolds (Mathematics) | Mathematical physics | AMS: Quantum theory – Quantum field theory; related classical field theories – Topological field theories | Manifolds and cell complexes – Differential topology – Topological quantum field theories

Classification: LCC QC20.7.T65 F834 2026

LC record available at https://lccn.loc.gov/2026002056

University Lecture Series ISSN: 1047-3998 (print); 2376-919X (online)

DOI: https://doi.org/10.1090/ulect/80

---

10 9 8 7 6 5 4 3 2 1   31 30 29 28 27 26

Contents

Preface

This book is intended to serve as a first introduction to topological field theories. As the term *topological field theory*, and even more so the synonymously used *topological quantum field theory* might suggest, the subject has its historical roots within the realm of physics. But the original physical ideas were picked up by mathematicians and gave rise to a mathematically precise notion of a topological field theory. Research concerned with topological field theories now lies at the crossroads of algebra, representation theory, topology, and mathematical physics.

A basic input in any topological field theory is a class of manifolds having some fixed dimension d. In this book our focus is on low-dimensional cases, in particular on three-dimensional theories. Understanding this case is an indispensable prerequisite as well as an excellent starting point for understanding more recent developments which involve theories in four and higher dimensions. Further, in order to make pertinent aspects of topological field theories fully concrete, we discuss in detail a class of topological field theories that are built from finite groups. These so-called Dijkgraaf-Witten theories allow for explicit computations entirely based on algebraic tools. Accordingly, no analytic constructions appear in this text.

The proper language handling the algebraic tools is supplied by the theory of tensor categories. Our introduction does not assume that the reader is familiar with this language. Rather, in the first two chapters of the book we provide, among other things, all relevant category-theoretic background.

This book aims at being accessible to readers from various different backgrounds, and also at different levels of education. It provides the fundamentals of topological field theories and gives a quick access to the literature on more challenging issues and recent developments. Specifically, we hope that an advanced undergraduate student in mathematics, or in physics with a solid mathematical background, will be able to use these notes for self-study. It would, however, be deceptive to assume that the text is customized to the preexisting knowledge of every single interested reader. We try to compensate this deficit by providing ample references to further background material and complementary literature.

As prerequisites we presume that the reader has acquired a certain degree of mathematical maturity, for instance by having been exposed to basic mathematical notions that are standard in the undergraduate curriculum of mathematics students and of many physics students. This comprises in particular the notion of a group, including the description of a group by generators and relations, the notion of a vector space, including the concept of a freely generated vector space and of the tensor product of vector spaces, and the one of an associative algebra A over a field, including A-modules, A-bimodules, and their tensor product. Textbooks covering this material are, e.g., [**Lan**] and [**DuF**].

We also expect a basic familiarity with a few aspects of differential topology, like the notion of a smooth manifold, and some intuition for compact oriented two-dimensional manifolds. As is apparent from the term *topological field theory*, some prerequisites from algebraic topology are needed as well. Apart from elementary concepts in point-set topology, we also assume a first exposure to covering theory, including in particular the notions of a fundamental group, of a universal cover and of a G-cover, as discussed, e.g., in Chapter 1.3 of [**Hat**].

On the other hand, no background in physics is formally required. Still, for appreciating some of our motivating considerations (which can, however, be skipped in a first reading), an acquaintance with basic concepts of quantum mechanics is of avail. The book [**FadY**], for instance, contains much more information than needed.

Acknowledgments

We are grateful to Azat Gainutdinov, Hannes Knötzele, Simon Lentner, Lukas Müller, David Reutter, Yorck Sommerhäuser, Max-Niklas Steffen, and Yilong Wang for valuable comments on the manuscript. We also thank the anonymous referees for their suggestions for improving the presentation. Jürgen Fuchs is supported by the Swedish Research Council (VR) under project no. 2022-02931. Christoph Schweigert is supported by the Deutsche Forschungsgemeinschaft (DFG) by SFB 1624 and under Germany's Excellence Strategy - EXC 2121 "Quantum Universe" – 390833306. Lukas Woike was supported by the Danish National Research Foundation through the Copenhagen Centre for Geometry and Topology (DNRF151) at Københavns Universitet during a period in which this book was written; he is also supported by the ANR project CPJ n° ANR-22-CPJ1-0001-01 at the Institut de Mathématiques de Bourgogne (IMB). The IMB receives support from the EIPHI Graduate School (contract ANR-17-EURE-0002).

An invitation to topological field theories

The term *topological field theory* has its historical roots in physics. As such, it originally involved a certain amount of heuristics motivated by general ideas from the realm of quantum field theory. Pertinent heuristic ideas originate from the study of d-dimensional quantum field theories on d-manifolds of the form $\Sigma \times I$, with Σ a $(d-1)$-dimensional manifold and I an interval corresponding to a "time direction". Adopting the paradigm of quantum mechanics, one wants to assign to Σ a vector space V_Σ of states, and to $\Sigma \times I$ an endomorphism of V_Σ that describes a time evolution. Requiring the theory to be topological amounts to demanding that this time evolution does in fact not depend on the choice of interval I; it may then be taken to be the identity linear map.

While this description is overly simplistic, it still captures the idea that topological quantum field theories embody ground states of quantum systems. Moreover, inspired by the topological notion of a *cobordism*, it is a fruitful idea to admit in place of the cylinders $\Sigma \times I$ more general d-manifolds M with boundary ∂M that are endowed with an orientation preserving diffeomorphism

$$\overline{\Sigma} \sqcup \Sigma \longrightarrow \partial M \,,$$

where the overline denotes orientation reversal. We then think of the manifold M as interpolating from Σ to Σ and write it as $\Sigma \xrightarrow{M} \Sigma$. This should actually be further generalized to d-manifolds M equipped with an orientation preserving map

$$\overline{\Sigma_1} \sqcup \Sigma_2 \longrightarrow \partial M \,.$$

Technically, this is a cobordism $\Sigma_1 \xrightarrow{M} \Sigma_2$; to such a cobordism we assign a linear map $V_M \colon V_{\Sigma_1} \longrightarrow V_{\Sigma_2}$. Thereby we deal with a two-layered structure that maps $d-1$-dimensional manifolds to vector spaces and cobordisms between them to linear maps. We think of this structure as a quantum system with a generalized time evolution.

A lot of compatibility conditions should be imposed on such structures: gluing of cobordisms should correspond to the composition of time evolutions, disjoint unions of manifolds should provide composite systems with composite evolutions, such composite systems should be well-behaved under permutation of its constituents, and so on. This large amount of data and of conditions calls for an efficient organization. The appropriate tool for achieving this are categories (and higher categories). Indeed, the definition of (the simplest incarnation of) a topological field theory reads, in its most compact form: a topological field theory of dimension d is a *symmetric monoidal functor from the category of d-dimensional cobordisms to the category of vector spaces.*

One goal of this book is to make the ingredients of this definition accessible. To this end we will elaborate, in Chapters 1 and 2, the category theoretic background in considerable depth. However, the basic idea behind the notion of a category can easily be explained without diving into technical details. In a nutshell, a *category* embodies a two-layered structure such as the one described above. Similar two-layered structures, and thus categories, arise in various contexts within mathematics and physics. Besides $d-1$-dimensional manifolds and d-dimensional cobordisms, other prototypical examples include: sets and maps between sets; vector spaces and linear maps; and manifolds and diffeomorphisms. Maps between such structures comprise two layers as well; they are captured by the notion of a *functor* between categories.

Already from this preliminary exposition it is clear that topological field theory involves both geometric and algebraic structures. Other mathematical disciplines are in fact relevant as well. Indeed, research concerned with topological field theories lies at the crossroads of algebra, representation theory, topology, and mathematical physics. Such research has been pursued intensively over the last three decades. It has steadily stimulated novel developments in these disciplines and created further links between them.

We have tried to account for the diverse motivations to study topological field theories and thereby make this book a useful resource for a broad readership. Let us briefly indicate some possible specific motivations for investigating topological field theories.

Motivation: Manifolds. Together with manifolds, one naturally also considers maps between them which preserve the manifold structure as well as further structures that the manifolds might be endowed with, such as an orientation. A different possibility for relating manifolds Σ and Σ' of the same dimension $d-1$ is the one already considered above, namely to specify a cobordism between them, that is, roughly, a d-dimensional manifold whose boundary consists of Σ and Σ'. Manifolds of a given dimension and cobordisms between them furnish the two layers of the *category of d-dimensional cobordisms*, which features as the domain of a topological field theory.

Accordingly, for the geometrically inclined mathematician topological field theories are of interest because they enable them to study manifolds in terms of linear objects such as vector spaces or, in other words, furnish a *linear representation* of the nonlinear structure of a manifold. This point of view is reinforced by the widespread experience that nonlinear algebraic structures, like finite groups or Lie groups, are efficiently studied through their linear representations, i.e., via vector spaces on which the structure acts.

Motivation: Manifold invariants. A topological field theory assigns in particular a number to every manifold of a given dimension d (and belonging to some specified class, say, closed oriented manifolds). Now whenever one encounters a numerical *invariant* of manifolds, it is a valid question to ask whether the invariant is naturally a part of a topological field theory. Such a topological field theory can then, for instance, explain in what sense the manifold invariant is *local*, i.e., can be computed entirely by considering smaller constituents into which the d-dimensional manifold can be cut up.

In the first step of such a process of chopping, d-dimensional manifolds with boundary arise. Accordingly a topological field theory assigns an algebraic quantity also to the $d-1$-dimensional boundary; in the simplest setting, this quantity is a vector space. As we will see, the definition of a topological field theory directly implies that the vector space assigned to a surface Σ comes with a representation of the group of isotopy classes of orientation preserving self-diffeomorphisms of Σ, i.e., of the *mapping class group* (see Remark 2.104). These groups are highly interesting objects at the interface of algebra and geometry.

In a further step, the manifolds with boundary that arise by cutting up d-manifolds can be chopped into even smaller pieces, leading to manifolds with corners. To express the corresponding additional amount of locality, a next entry in the hierarchy of objects that starts with numbers and vector spaces is needed, and indeed at this point linear categories naturally enter the game. In this way, a topological field theory provides, in fact, an additional third layer of structure.

Motivation: Higher algebraic structures. The flavor of the algebraic structures that a topological field theory assigns to geometric objects is strongly correlated with the dimension of the topological field theory. Specifically, one finds that one-dimensional topological field theories are closely related to finite-dimensional linear algebra, while the "higher" algebraic structures provided by algebras and their modules appear in two-dimensional topological field theories. In three-dimensional topological field theories one encounters an even higher structure: tensor categories.

Important classes of tensor categories arise, in turn, as representation categories of suitable algebraic structures, like vertex algebras (which, e.g., encode chiral symmetries of two-dimensional conformal field theories), Hopf algebras, and quantum groups. The study of such structures is an active field of research in representation theory, with an impressive track record. Topological field theories provide expedient tools for organizing and visualizing salient features of these structures by means of topology. Furthermore, algebras, bi- and Hopf algebras, as well as Frobenius algebras can not only be defined as structures on vector spaces, but also on objects of more general tensor categories. In short, for the algebraist, topological field theory furnishes a well-motivated organization principle and guide to higher structures in algebra.

Motivation: Quantum systems. Let us also be more specific about the provenance of topological field theories in physics. One of the roots are models for three-dimensional quantum gravity that make use of discretizations of three-manifolds. These models are the origin of *state-sum constructions* of topological field theories.

Another source is the field-theoretic description of quantum Hall systems. This involves the issue of cancellation of anomalies, which necessitates the introduction of topological terms in the action functional that describes the three-dimensional bulk system. A prominent example for an action of this type is the Chern-Simons action. A heuristic analysis of the corresponding quantum field theory has been the basis for another construction of topological field theories which makes use of the surgery of manifolds.

The evaluation of an extended three-dimensional topological field theory on a circle yields the structure of a *modular fusion category*, also known as semisimple modular tensor category. Modular fusion categories arise in many different situations. For instance, they can be used for characterizing topologically ordered phases of matter. In this application, they capture in particular information on types of quasi-particle excitations, such as their braid group statistics. Specifically, in this context a simple object in a modular fusion category describes a type of *anyon*.

Why categories? The considerations above highlight that category theory plays a crucial role in the investigation of topological field theories. But do they even call for formulating the very definition of a topological field theory entirely in category-theoretic terms? Admittedly, there exist approaches to topological field theories that avoid such a formulation, e.g., descriptions in terms of action functionals and path integrals. But, apart from possible issues concerning mathematical rigor, even such approaches tend to give rise to quantities that are naturally organized in categorical terms. For example, even when formulated heuristically with an action and a path integral, a Chern-Simons theory gives rise to a *category* of Wilson lines. Indeed, in any setting that covers arbitrary types of topological field theories, category-theoretic concepts play a critical role. This is the framework adopted in the present book. Indeed we think that our exposition demonstrates abundantly that the use of categories is indispensable for a thorough understanding of topological field theories. Conversely, topological field theories provide, in our opinion, a particularly well-motivated access to category theory.

Actually, categories, in the form of *two*-layered structures, are not the end of the story. We have already seen that the process of chopping manifolds into pieces, which results in locality properties, can be iterated. This leads to additional layers of structure. The same happens in many other contexts. For instance, when considering the two-layered structure provided by algebras and bimodules over them, a natural third layer is provided by maps between bimodules. The proper framework for dealing with such situations is the one of *higher categories*, and specifically in the case of three layers, of *bicategories*. There are then analogous additional layers for functors between higher categories and hence, as a specific instance, for topological field theories. These higher variants are often called *extended topological field theories*. We will study such extended theories in Chapter 5, where we also provide a brief introduction to bicategories.

From a more general perspective, it is worth pointing out the following aspects of categories and topological field theories that are of relevance for mathematicians and physicists alike: On the one hand, categories provide a powerful unifying language for addressing mathematical structures; on the other hand, the study of topological field theories serves as a driving force in the development of higher categorical concepts. The efficiency of category-theoretic language is reflected by the ever increasing use of categories, as well as of higher categories, in mathematical physics.

Dijkgraaf-Witten theories. For making various aspects of topological field theories as concrete as possible, we concentrate our attention on a particular class of topological field theories that are formulated in geometric terms, the so-called *Dijkgraaf-Witten theories*. Dijkgraaf-Witten theories can be constructed with the help of basic concepts from topology and linear algebra. As a consequence, as we try to exhibit in this book, they allow for explicit computations. For instance, the category that an extended three-dimensional Dijkgraaf-Witten theory assigns to a circle can be determined, and a balanced braided structure for this category can be directly read off from the geometry; see Section 5.5.

At the same time, Dijkgraaf-Witten theories illustrate fundamental principles of the physics concept of *gauge theories*. The term "gauge theory" indicates the presence of a gauge symmetry. Such a symmetry is really an overparametrization of the configurations of the physical system. In particular, all observables take the same value on gauge equivalent configurations. (This notion of a gauge theory comprises the one of gauge theories formulated in terms of a Lagrangian, as appearing, e.g., in particle physics, but is considerably more general.)

The gauge symmetries in a Dijkgraaf-Witten theory form a finite group. This allows us to study them entirely within the realm of algebra. Accordingly, no analytic constructions will appear in the text. Another crucial feature of these quantum field theories is that the field configurations are not to be considered as elements of a set, but rather, in order to preserve locality properties, as objects in a category (in fact, a *groupoid*). Indeed it will become clear in our discussion of Dijkgraaf-Witten theories that category theory supplies the appropriate tools for dealing with the overparametrization that is present in gauge theories.

Outline. We end this invitational section with a brief survey of the contents of the book. As already mentioned, most of the background from category theory is treated in Chapters 1 and 2. We proceed in a step-by-step fashion: basic notions as well as examples of categories are provided in Sections 1.1 and 1.2 and functors are discussed in Section 1.3, while Sections 1.5 and 1.6 are devoted to the more advanced topics of *limits* and *colimits* and of *adjointness* of functors, respectively. Chapter 2 deals with additional structures and properties that categories may possess and that are present both for the category of cobordisms and for the category of vector spaces: *monoidal* structures in Sections 2.2 and 2.4, *rigid dualities* in Section 2.5, and *braidings* in Section 2.7. In addition, pertinent aspects of bicategories, as needed for the discussion of extended topological field theories, are covered in Section 5.1.

When working with categorical structures, a comfortable tool is a graphical calculus in terms of so-called *string diagrams*. We explain this calculus, again step by step, in specific parts earmarked as *Graphical Description*; they are numbered as 1.16, 2.17, 2.60, and 2.94, as well as 5.9 for the case of bicategories.

Apart from the category-theoretic input, Chapters 1 and 2 also introduce notions and constructions from geometry and algebra which are needed in our exposition of topological field theories. On the geometry side, this includes, e.g., in Section 1.2 the concepts of fundamental groupoid, principal bundle, and cobordism, and in Section 2.1 the one of a classifying space, as well as the description of the rigid duality for cobordisms in Section 2.5. On the algebra side, most of the Sections 2.3 and 2.6 is devoted to aspects of bialgebras and of Hopf algebras, respectively, and we also exhibit constructions like the tensor product of bimodules (Section 1.5) and the extension of scalars (Section 1.6). The algebraic input is again amply backed up with string diagrams; see the Graphical Descriptions 2.22, 2.30, 2.82, and 2.85.

Yet another topic treated in Chapter 2 is a motivation of the functorial definition of topological field theories from physical principles (Section 2.1). This comprises in particular two parts which summarize some basic principles of quantum field theories, including the notion of field configurations and of gauge invariance of the action functional (Principles of field theory 2.1), and of boundary values, state space, and transition amplitudes (Principles of field theory 2.4).

In our discussion of topological field theories we focus on low-dimensional theories. Understanding these cases, and in particular the one of dimension three, is still an indispensable prerequisite and an excellent starting point for dealing with more recent developments. Chapter 3 starts this subject by considering theories in dimension one and two. It addresses, e.g., the issue of *classification*, in the sense of understanding a list of items of interest by mapping it to a different list of items of an already more familiar type. The classification of topological field theories of a given dimension is intimately related to the structure of all manifolds of that dimension. As a first embodiment of this connection, we exhibit in Section 3.1 that one-dimensional topological field theories are classified by finite-dimensional vector spaces (see Theorem 3.1). In the other sections of Chapter 3 we use the structure of compact oriented two-manifolds to present the classification of two-dimensional oriented topological field theories by commutative Frobenius algebras (Theorem 3.14). Pertinent information about the latter type of algebras is provided in Section 3.3.

In Chapter 4 we concentrate our attention on Dijkgraaf-Witten topological field theories. The definition of these theories as a specific functor from cobordisms to vector spaces is given in Section 4.2 (Definition 4.37). To show that they indeed constitute topological field theories requires some effort; it is finally achieved in Section 4.4 (Theorem 4.66). In Chapter 5 we proceed to the realm of extended topological field theories. This includes in particular, in Sections 5.2 and 5.3, an exposition of three-dimensional Dijkgraaf-Witten theories, which are constructed in a two-step process (see Definition 5.25). This chapter culminates in the classification of 3-2-1-extended three-dimensional topological field theories in terms of modular fusion categories (Theorem 5.60). The notion of modular category which crucially enters this classification is provided in Definitions 5.49 and 5.53.

The final Chapter 6 has a different flavor. It provides an outline of a biased selection of topics of recent and ongoing interest. These are largely intended as appetizers and as an invitation for further reading. We refrain from trying to list all topics taken up and only mention the following issues. First, concrete constructions of topological field theories are typically based on suitable algebraic input data. As examples we mention, in Section 6.4, the string-net construction and state-sum models, for which the required input is captured by a pivotal fusion category, as well as the surgery construction, which takes a modular fusion category as input. Second, topological field theories can be used as tools for organizing and visualizing features of various algebraic structures; we present a few instances of this phenomenon in Section 6.5. And third, in Section 6.3.2 we offer an outlook on one major insight in the field, the *cobordism hypothesis*, which comprises a statement about the structure of all d-dimensional manifolds for each value of d. We hope that by studying our introductory notes on topological field theories the reader will be prepared and motivated to learn more about this and other advanced topics.

Literature. Let us finally include a few pointers to the literature. Just like the choice of topics of the book, this list is biased by our taste and the limits of our knowledge. Moreover, we do not aspire to give appropriate credit to the original literature, but instead favor expository sources that are of avail for the intended readership.

We are confident that our exposition of the category-theoretic background in Chapters 1 and 2 and in Chapter 5.1 covers all that is needed for the purposes of this book. Still, our text is undoubtedly not intended to provide a stand-alone introduction to category theory. Readers looking for such an introduction may consult a textbook like [**Ri2**] or [**AdHS**], and are encouraged to read parts of [**Ri2**] in parallel where indicated. A standard reference for monoidal categories and monoidal categories with additional structure such as a braiding, as well as for structures related to module categories, is [**EtGNO**].

Pertinent information on topological field theories can, e.g., be found in [**TV1**] and for two-dimensional theories in [**Koc**]. The original article on the construction of topological field theories via surgery [**Wit**] fits well as a complementary reading. For further background on bi- and Hopf algebras we refer to [**Kas**, **Kash**], and for an introduction to vertex algebras to [**FreB**] (see also [**FSWY**] for a brief summary).

More specific references to the literature are given at appropriate places in the text.

CHAPTER 1

Categories

As pointed out in the Preface, categories, as well as concepts generalizing categories, provide a powerful unifying language for addressing mathematical structures, including in particular crucial ingredients of topological field theories. Accordingly, in this first chapter we set up the basic notions from category theory that will be indispensable tools in our discussion of topological field theories.

One specific ingredient in our constructions will be vector spaces over a field $\mathbb{K}$. Throughout these notes, $\mathbb{K}$ is assumed to be algebraically closed. In applications in physics, $\mathbb{K}$ is usually the field $\mathbb{C}$ of complex numbers.

1.1. Categories: Definition and examples

To get started, let us consider one of the arguably simplest geometric systems.

EXAMPLE 1.1. A compact oriented zero-dimensional manifold Y is a finite set of points, with each point either positively oriented (to be denoted by pt_+) or negatively oriented (denoted by pt_-). Disjoint union provides a binary multiplication of such sets. We also admit the empty set as an oriented zero-dimensional manifold; it behaves as a unit under the binary multiplication.

We may relate two compact oriented zero-dimensional manifolds Y_1 and Y_2 by any configuration of finitely many closed oriented circles and arrows, i.e., oriented intervals, subject to the condition that the starting point of an arrow is either a positively oriented point on the 'incoming' manifold Y_1 or a negatively oriented point on the 'outgoing' manifold Y_2, while allowed endpoints of arrows are negatively oriented points on Y_1 and positively oriented points on Y_2. An example of such a configuration, which should convey the idea, is shown in the following picture:

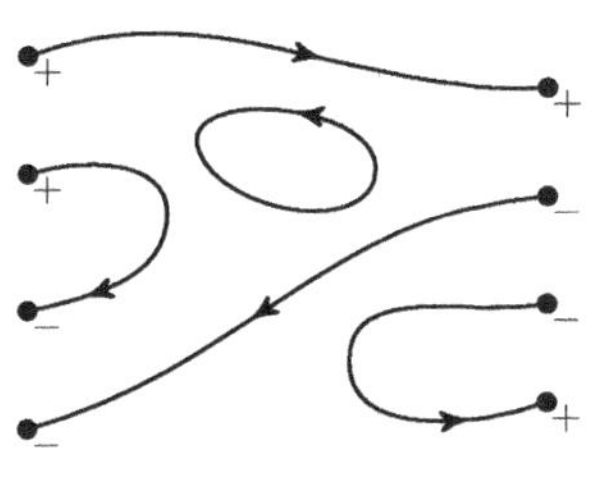

Two configurations can be concatenated, provided that the outgoing manifold of the first configuration coincides with the incoming manifold of the second configuration.

Here is an example of such a concatenation.

(1.2) 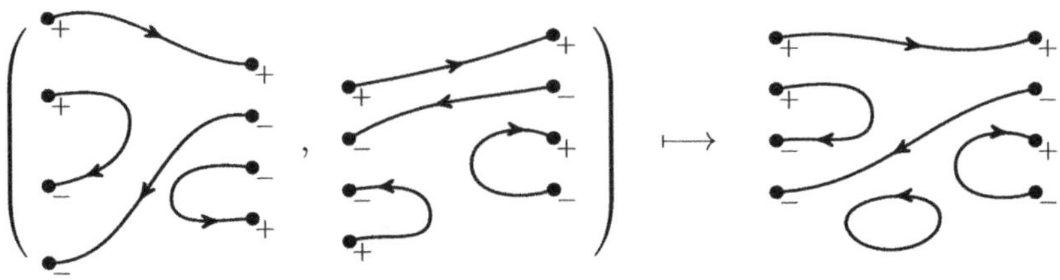

Note that we do not assign a length to an arrow or any of the circles and allow for small local deformations. As a consequence, the concatenation of two configurations of circles and arrows (if it exists) is again an allowed configuration. Moreover, if the incoming and outgoing manifold are the same, $Y_1 = Y_2$, then there is a distinguished configuration in which the arrows go from a point in Y_1 to the corresponding point in Y_2, as illustrated in the following picture.

(1.3) 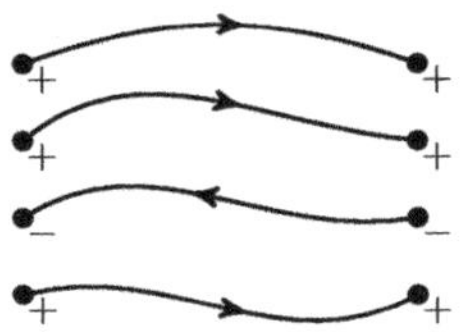

This way we end up with a two-layered structure, compact oriented zero-dimensional manifolds and configurations of oriented intervals and circles. To each configuration we can assign an incoming and an outgoing manifold. Also, there is a partially defined composition of configurations; this composition is associative, and the distinguished configurations of the type just shown act as identities under composition.

Two-layered structures analogous to the one observed in Example 1.1 are in fact very common. Abstracting from the examples one introduces the following notion:

DEFINITION 1.4. A *category* $\mathcal{C}$ consists of the following data:

(1) a class $\mathrm{Obj}(\mathcal{C})$, whose elements are called the *objects* of $\mathcal{C}$;
(2) a class $\mathrm{Hom}(\mathcal{C})$, whose elements are called *morphisms* of $\mathcal{C}$;
(3) four maps

$$\begin{aligned} \mathrm{id}: &\quad \mathrm{Obj}(\mathcal{C}) \longrightarrow \mathrm{Hom}(\mathcal{C}), \\ s,t: &\quad \mathrm{Hom}(\mathcal{C}) \longrightarrow \mathrm{Obj}(\mathcal{C}), \\ \circ: &\quad \mathrm{Hom}(\mathcal{C}) \times_{\mathrm{Obj}(\mathcal{C})} \mathrm{Hom}(\mathcal{C}) \longrightarrow \mathrm{Hom}(\mathcal{C}). \end{aligned}$$

Here $\mathrm{Hom}(\mathcal{C}) \times_{\mathrm{Obj}(\mathcal{C})} \mathrm{Hom}(\mathcal{C})$ consists by definition of those pairs of morphisms (g, f) in $\mathcal{C}$ which are *composable*, i.e., obey $t(f) = s(g)$.

We write

$$\mathrm{Hom}_{\mathcal{C}}(V, W) := \{f \in \mathrm{Hom}(\mathcal{C}) \mid s(f) = V \text{ and } t(f) = W\}$$

for $V, W \in \mathrm{Obj}(\mathcal{C})$.

For any pair V, W of objects we require $\mathrm{Hom}_{\mathcal{C}}(V, W)$ to be a set.

We call $\mathrm{id}(V)$ the *identity morphism* on $V \in \mathrm{Obj}(\mathcal{C})$, $s(f)$, and $t(f)$ the *source* (or *domain*) and *target* (or *codomain*) of $f \in \mathrm{Hom}(\mathcal{C})$, and $\circ$ the *composition* of morphisms. Instead of $\mathrm{id}(V)$ we will denote the identity morphism on V also by id_V; another common notation is to write just V for it, i.e., to tacitly identify an object and its identity morphism. For the composition of a pair $(g, f) \in \mathrm{Hom}(\mathcal{C}) \times_{\mathrm{Obj}(\mathcal{C})} \mathrm{Hom}(\mathcal{C})$ of morphisms we also employ the notation $\circ(g,f) =: g \circ f$. The four maps id, s, t and $\circ$ are subject to the following relations:

(1) $s(\mathrm{id}_V) = V = t(\mathrm{id}_V)$ for every $V \in \mathrm{Obj}(\mathcal{C})$;
(2) $\mathrm{id}_{t(f)} \circ f = f = f \circ \mathrm{id}_{s(f)}$ for every $f \in \mathrm{Hom}(\mathcal{C})$;
(3) the associativity identity

$$(h \circ g) \circ f = h \circ (g \circ f)$$

for every triple of morphisms $f, g, h \in \mathrm{Hom}(\mathcal{C})$ with $t(f) = s(g)$ and $t(g) = s(h)$.

The notion of a category can also be motivated in various other ways. We refrain from presenting any particular of them. Instead we present a list of specific categories which should suffice to illustrate the utility of this concept:

EXAMPLES 1.5. Here are a few common examples of categories:

(1) Any set S can be endowed with a trivial structure of a category for which the objects are the elements of S and the only morphisms are the identity morphisms.
Such a category is called a *discrete category*.
(2) Sets together with maps of sets form a category $\mathcal{S}et$. The identity morphism of any set is the identity map, and composition of morphisms is composition of maps.
As we will see in the next examples, there are many types of categories whose objects are sets equipped with some additional structure and whose morphisms are structure-preserving maps. This justifies the arrow notation $V \xrightarrow{f} W$ for morphisms; it also leads to the use of the term *map*, or also *arrow*, as a synonym for morphism.
(3) Given a field $\mathbb{K}$, the vector spaces over $\mathbb{K}$ as objects, together with linear maps between them as morphisms, form a category $\mathcal{V}ect(\mathbb{K})$.
A particular feature of this category is that its morphism sets are not merely sets, but even $\mathbb{K}$-vector spaces, and that composition is $\mathbb{K}$-bilinear. Any category sharing these specific features is said to be *enriched* over the category $\mathcal{V}ect(\mathbb{K})$ or, equivalently, to be $\mathbb{K}$-*linear* (or also just *linear*, when the field $\mathbb{K}$ in question is obvious). $\mathbb{K}$ is then also called the *ground field* of the category. For an actual definition of the general notion of enriched category we refer to Chapter 3 of [**Ri1**].
(4) More generally, given a ring R, the (left or right) modules over R together with R-module morphisms (also called R-module maps) form a category R-mod.
Note that in the case that $R = \mathbb{Z}$ is the ring of integers, an R-module is the same as an abelian group.
(5) As a generalization of $\mathcal{V}ect(\mathbb{K})$, for any group G there is the category $\mathcal{V}ect_G(\mathbb{K})$ of G-graded $\mathbb{K}$-vector spaces.
The objects of $\mathcal{V}ect_G(\mathbb{K})$ are $\mathbb{K}$-vector spaces that have a direct sum decomposition $V = \bigoplus_{g\in G} V_g$; the elements of the subspace V_g are called *homogeneous* elements of degree g. The morphisms are $\mathbb{K}$-linear maps that preserve the grading.

(6) Consider a category with a single object, which we denote by the symbol '$*$'. This category is completely described by the monoid $\mathrm{End}(*)$.
If we pass to the enriched setting, then one-object categories also allow us to describe algebras, i.e., monoids in vector spaces: We can consider for any associative unital $\mathbb{K}$-algebra A the category $*//A$ with a single object and morphisms $\mathrm{End}(*)$ given by the algebra A. This category is enriched over the category of $\mathbb{K}$-vector spaces.
(7) The category $\mathcal{T}op$ has as objects topological spaces and as morphisms continuous maps of topological spaces.
(8) The category $\mathcal{M}an$ has as objects smooth finite-dimensional manifolds and as morphisms smooth maps of manifolds.
Throughout this book all manifolds will tacitly be assumed to be smooth and finite-dimensional.

The morphisms of a category need not be maps of any sort. We have in fact already encountered a category for which this is the case, namely in the situation considered in Example 1.1:

EXAMPLE 1.6. The *cobordism category* $\mathcal{C}ob_{1,0}$ is the following category: An object in $\mathcal{C}ob_{1,0}$ is a compact oriented zero-dimensional manifold, i.e., a disjoint union of finitely many oriented points

$$\mathrm{pt}_+ = (\bullet, +) \qquad \text{and} \qquad \mathrm{pt}_- = (\bullet, -)\,.$$

A morphism in $\mathcal{C}ob_{1,0}$ is an oriented one-dimensional manifold, i.e., a disjoint union of intervals and circles, up to diffeomorphism. Composition is concatenation of representative one-manifolds and identity morphisms are (classes of) disjoint unions of intervals, as illustrated in the pictures (1.2) and (1.3), respectively. The category $\mathcal{C}ob_{1,0}$ comes with additional structure. We are not yet in a position to formulate that structure, though, since this will require quite a few more mathematical notions. Once we have supplied these ingredients, we will provide, in Definition 1.36, a precise description of $\mathcal{C}ob_{1,0}$, as well as of its higher-dimensional analogs.

DEFINITION 1.7.

(1) A *subcategory* $\mathcal{S}$ of a category $\mathcal{C}$ is a subcollection of the objects and morphisms of $\mathcal{C}$ such that for any morphism $f\colon X \to Y$ in $\mathcal{S}$ the objects $X, Y \in \mathcal{C}$ are in $\mathcal{S}$, the composite of morphisms in $\mathcal{S}$ is again in $\mathcal{S}$, and $\mathcal{S}$ contains the identity morphism in $\mathcal{C}$ of any object in $\mathcal{S}$.
(2) A *full subcategory* of $\mathcal{C}$ is a subcategory $\mathcal{S}$ such that for every pair of objects $X, Y \in \mathcal{S}$ one has $\mathrm{Hom}_{\mathcal{S}}(X,Y) = \mathrm{Hom}_{\mathcal{C}}(X,Y)$.

EXAMPLES 1.8.

(1) The category of finite sets, with morphisms given by maps between sets, is a full subcategory of the category of all sets.
(2) The category of abelian groups is a full subcategory of the category of groups.
(3) The category of vector spaces over a field $\mathbb{K}$, with morphisms given by $\mathbb{K}$-linear maps, is a subcategory of the category of sets that is not full.

REMARK 1.9. For the reader interested in set-theoretic issues (others may wish to skip this Remark) we outline why the definition of a category involves *sets* as well as *classes*. Namely, for applying category theory in practice one would like to have the notion of a "category of all sets" and, for constructing interesting categories, for a given property $\varphi(x)$ of a set x, also a subcategory "$\{x \mid \varphi(x)\}$" of all sets having the property φ. Famously, this leads to paradoxa, such as the following, which is known as Russell's antinomy: Consider the putative set $R := \{M \mid M \notin M\}$ of all sets that are not elements of themselves; if R existed, then one would have to conclude that $R \in R$ if and only if $R \notin R$.

A resolution of this problem is to restrict the application of φ to sets that are elements in some specific set $\mathfrak{U}$ (where it is supposed that the notion of a set is defined, e.g., by working with Zermelo-Fraenkel axioms). Further, such a set $\mathfrak{U}$ must be sufficiently nice – technically speaking, it must be a *universe* (for details see [**Mac**, Sect. I.6]). All mathematical constructions are then carried out inside the universe $\mathfrak{U}$. A set that is an element of $\mathfrak{U}$ is called *small* (relative to $\mathfrak{U}$). It should be appreciated that, with this definition, sets that are small in terms of cardinality are not necessarily $\mathfrak{U}$-small; for example, the one-element set $\{\mathfrak{U}\}$ is *not* $\mathfrak{U}$-small. Functions between small sets relative to $\mathfrak{U}$ can be constructed inside $\mathfrak{U}$. This yields a category of $\mathfrak{U}$-small sets.

A category $\mathcal{C}$ is now called $\mathfrak{U}$-*small* if the set $\mathrm{Obj}(\mathcal{C})$ of objects is in $\mathfrak{U}$. The category of $\mathfrak{U}$-small categories is *not* $\mathfrak{U}$-small, because this would imply $\mathfrak{U} \in \mathfrak{U}$, violating the axioms of a universe. A *class* C (relative to a universe $\mathfrak{U}$) can then be defined as an arbitrary subset $C \subseteq \mathfrak{U}$. It follows that every $\mathfrak{U}$-small set is a $\mathfrak{U}$-class, but the converse is not true. Using classes, we can now talk about the category of $\mathfrak{U}$-small categories.

The choice of $\mathfrak{U}$ is usually suppressed in the notation, and we will follow this habit in this book. It is common to enlarge the axioms of set theory by requiring that for any set X there is a universe $\mathfrak{U}$ such that $X \in \mathfrak{U}$, which in particular ensures the existence of universes. Most of the categories appearing in this book are *essentially small*, meaning that they are *equivalent* to a small category. The relevant notion of equivalence of categories will be provided in Definition 1.60.

It is often convenient to formulate an equality between morphisms in the form of a *commutative diagram*. A diagram built from objects and morphisms between them is commutative if the composite morphisms formed by any two paths of composable morphisms that start at the same source object and end at the same target object are equal. For instance, that $g \circ f$ equals the composition of f and g amounts to the statement that the diagram

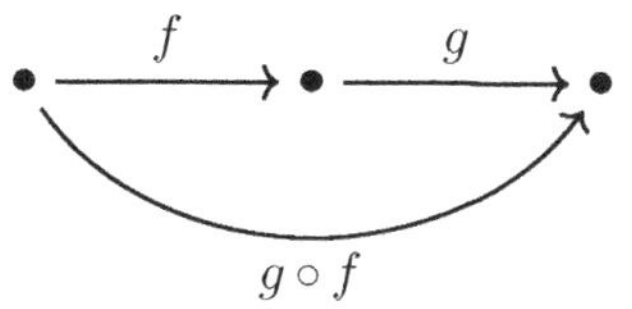

is commutative. Here we suppress the information about the objects involved by representing each of them by a black circle. With the same notation, associativity of composition of morphisms can be expressed as commutativity of a diagram of the following shape:

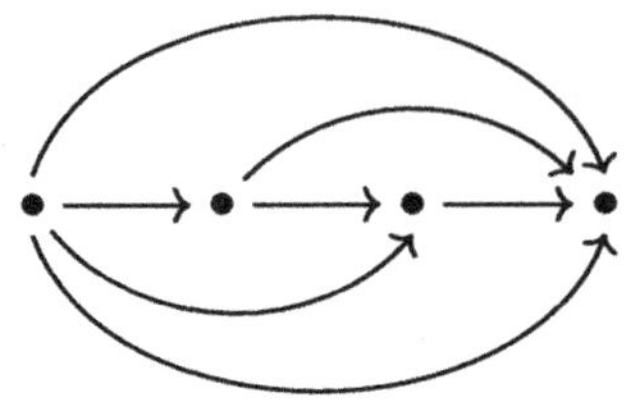

DEFINITION 1.10. Let $\mathcal{C}$ be a category.

(1) To indicate that a morphism $f \in \mathrm{Hom}(\mathcal{C})$ is an element of the set $\mathrm{Hom}(V, W)$ we use the notation $V \xrightarrow{f} W$ or, equivalently, $f\colon V \longrightarrow W$.

(2) A morphism $f\colon V \longrightarrow W$ is called an *isomorphism* if it has a two-sided inverse, i.e., if there exists a morphism $g\colon W \longrightarrow V$ such that

$$g \circ f = \mathrm{id}_V \quad \text{and} \quad f \circ g = \mathrm{id}_W .$$

Two objects V and W for which an isomorphism $f\colon V \longrightarrow W$ exists are called *isomorphic*. The class of isomorphism classes of objects of a category $\mathcal{C}$ is denoted by $\mathrm{Iso}(\mathcal{C})$. The morphisms f and g are also called *invertible* and one writes $g = f^{-1}$. We write $V \cong W$ to express the property that V and W are isomorphic, and also $f\colon V \xrightarrow{\cong} W$ to indicate the structure furnished by the choice of a specific isomorphism f.

(3) An element of $\mathrm{Hom}_{\mathcal{C}}(V, V) =: \mathrm{End}_{\mathcal{C}}(V)$ is called an *endomorphism* of V. An isomorphism in $\mathrm{End}_{\mathcal{C}}(V)$ is called an *automorphism* of V. The subset of automorphisms in $\mathrm{End}_{\mathcal{C}}(V)$ is denoted by $\mathrm{Aut}_{\mathcal{C}}(V)$.

(4) A morphism $f\colon X \longrightarrow Y$ in $\mathcal{C}$ is called a *monomorphism* if for any two morphisms $g, h\colon Z \longrightarrow X$ the equality $f \circ g = f \circ h$ implies that $g = h$.

(5) A morphism $f\colon X \longrightarrow Y$ in $\mathcal{C}$ is called an *epimorphism* if for any two morphisms $g, h\colon Y \longrightarrow Z$ the equality $g \circ f = h \circ f$ implies that $g = h$.

REMARKS 1.11.

(1) If a morphism f is invertible, then its inverse f^{-1} is unique.

(2) The endomorphisms of any object V form a monoid with respect to composition. This monoid is unital, with the identity morphism id_V of V as the unit.

(3) The automorphisms of any object V form a group. The set $\mathrm{Aut}_{\mathcal{C}}(V)$ is therefore also called the *automorphism group* of V.

REMARK 1.12. In category theory one often encounters situations in which it would not make sense to ask whether two given objects are *equal* on the nose. Instead, one should ask whether they are *isomorphic*.
For instance, any finite-dimensional vector space V is isomorphic to its dual vector space $V^* = \mathrm{Hom}_{\mathbb{K}}(V, \mathbb{K})$. But there is no distinguished such isomorphism, as is, e.g., seen by noticing that in order to exhibit an isomorphism one must choose bases of V and V^*.

As a more subtle example, consider complex finite-dimensional representations of the Lie group SU(2). We should better not ask whether such a representation – say, the defining two-dimensional representation – is equal to its dual. Once we ask instead whether it is isomorphic to its dual, we can pose refined questions about the isomorphism. This leads, e.g., to the distinction between real and quaternionic (also called pseudo-real) representations, i.e., representations endowed with a bilinear form that is symmetric and antisymmetric, respectively. Concretely, every finite-dimensional irreducible representation of SU(2) is isomorphic to its dual; the integer-spin representations are real, while the half-integer-spin representations are quaternionic.

DEFINITION 1.13.

(1) Given any category $\mathcal{C}$, its *opposite category* (sometimes also called *opposed* category) $\mathcal{C}^{\text{opp}}$ implements the idea of "reversing arrows": $\mathcal{C}^{\text{opp}}$ has the same objects as $\mathcal{C}$, while its morphisms are

$$\text{Hom}^{\text{opp}}(U,V) := \text{Hom}(V,U)\,.$$

Composition of morphisms in $\mathcal{C}^{\text{opp}}$ is defined by

$$f \circ^{\text{opp}} g := g \circ f\,,$$

with $\circ$ the composition in $\mathcal{C}$.

Together with any categorical concept there comes a second one, which is obtained by applying the former in the opposite setting, in which all arrows are reversed. One often refers to two statements that are related this way as being *dual* to each other.

(2) The *Cartesian product* $\mathcal{C}\times\mathcal{D}$ of two categories $\mathcal{C}$ and $\mathcal{D}$ is the following category: Objects in $\mathcal{C}\times\mathcal{D}$ are pairs (U,U') with U an object in $\mathcal{C}$ and U' an object in $\mathcal{D}$. A morphism $(U,U')\longrightarrow(V,V')$ in $\mathcal{C}\times\mathcal{D}$ is a pair (f,f'), where $f\colon U\longrightarrow V$ is a morphism in $\mathcal{C}$ and $f'\colon U'\longrightarrow V'$ is a morphism in $\mathcal{D}$. Composition of morphisms is given by $(g,g')\circ(f,f') := (g\circ f, g'\circ f')$.

(3) The *disjoint union* $\mathcal{C}\sqcup\mathcal{D}$ of two categories $\mathcal{C}$ and $\mathcal{D}$ is the category with objects $\text{Obj}(\mathcal{C}\sqcup\mathcal{D}) := \text{Obj}(\mathcal{C})\sqcup\text{Obj}(\mathcal{D})$ and with morphism sets

$$\text{Hom}_{\mathcal{C}\sqcup\mathcal{D}}(X,Y) := \begin{cases} \text{Hom}_{\mathcal{C}}(X,Y) & \text{for } X,Y\in\mathcal{C}\,,\\ \text{Hom}_{\mathcal{D}}(X,Y) & \text{for } X,Y\in\mathcal{D}\,,\\ \emptyset & \text{else}\,. \end{cases}$$

The composition is inherited from the compositions of $\mathcal{C}$ and $\mathcal{D}$.

For any group G we can study its representations on vector spaces over a field $\mathbb{K}$. In more detail, a $\mathbb{K}$-linear *representation* (V,ρ) of G is a $\mathbb{K}$-vector space V together with an invertible linear map $\rho(g)\in\text{GL}(V)$ for each group element $g\in G$, in a way compatible with the group multiplication, i.e., such that $\rho(g)\circ\rho(h)=\rho(gh)$ for all $g,h\in G$. Put differently, ρ is a group homomorphism from G to the group $\text{GL}(V)$ of invertible linear maps from V to itself. An *intertwiner*

$$\varphi\colon\quad (V,\rho_V)\longrightarrow(W,\rho_W)$$

is a linear map $\varphi\colon V\longrightarrow W$ that satisfies

$$\rho_W(g)\circ\varphi = \varphi\circ\rho_V(g)$$

for every $g\in G$.

EXAMPLE 1.14. Given a group G and a field $\mathbb{K}$, the representations of G on vector spaces over $\mathbb{K}$, together with intertwiners, form a category. We denote it by G-$\mathcal{R}ep$.

An alternative description of G-$\mathcal{R}ep$ is obtained via

DEFINITION 1.15. The *group algebra* $\mathbb{K}[G]$ of a group G is the $\mathbb{K}$-vector space with basis $\{\beta_g \,|\, g \in G\}$ and with associative multiplication given by the bilinear extension of

$$\beta_g \cdot \beta_h := \beta_{gh}$$

for $g, h \in G$.

A module over the group algebra $\mathbb{K}[G]$ is the same as a representation of G on a $\mathbb{K}$-vector space. Accordingly, the categories G-$\mathcal{R}ep$ can be seen as special instances of Example 1.5(4).

GRAPHICAL DESCRIPTION 1.16. A convenient way to think of morphisms of a category is in terms of a suitable graphical calculus, including the option to account for possible further structure on the category, specifically those that will be introduced in Chapter 2. Plenty of variants of such a calculus (as well as of conventions) are in use in the literature. A common theme is to think of objects X as labeling wires, or *strings*, and of a morphism from X to Y as a node at which an input string labeled X is turned into an output string labeled Y. Accordingly, the resulting graphs are often called *string diagrams.*

Our basic conventions for string diagrams are as follows. Strings labeled by objects are preferably drawn as straight vertical lines, and a node representing a morphism is depicted as a box, also called a *coupon*, with the input string attached to the bottom of the coupon and the output string attached to its top. To stress the distinction between in- and outputs, we draw coupons with a bottom that is a bit larger than their top. Thus a morphism f in $\mathrm{Hom}(X, Y)$ is depicted as the string diagram

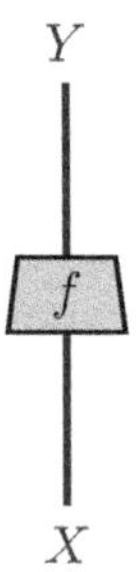

with the domain X of f at the bottom and the codomain Y at the top. Specifically, the diagram is to be read from bottom to top. In line with the alternative convention to write just X in place of id_X, in the particular case of the identity morphism on an object X the coupon can be omitted:

(1.17)

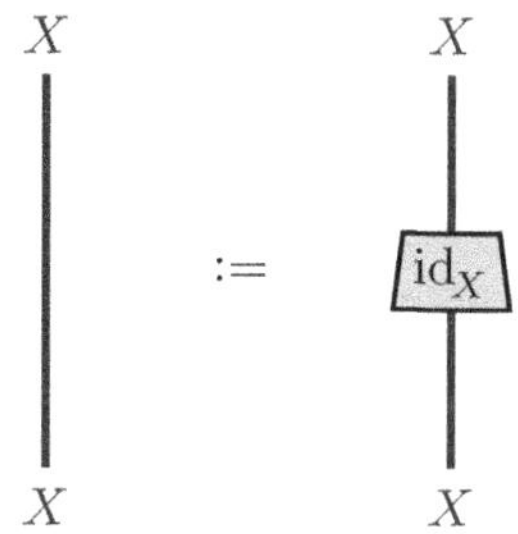

Up to this point the graphical calculus keeps track of the source and target maps of a category and of the identity morphisms, while it does not yet account for the composition of morphisms. The latter is implemented in the form of a move that modifies the graph locally: adjacent coupons in a string diagram describe composable morphisms, and the composition map $\mathrm{Hom}(Y,Z) \times \mathrm{Hom}(X,Y) \to \mathrm{Hom}(X,Z)$ can be used to replace a pair of adjacent coupons by a single coupon, according to

(1.18)

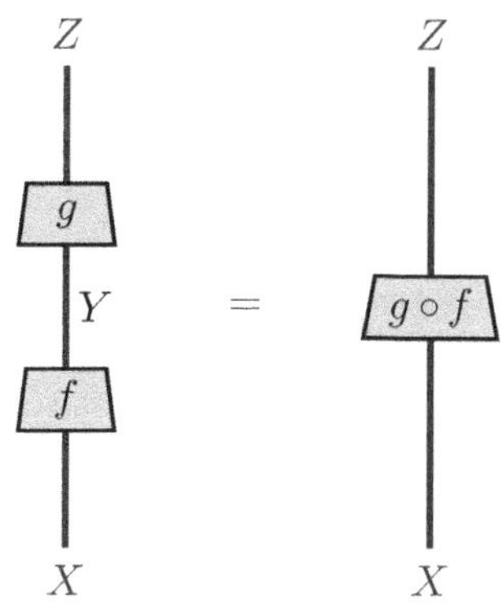

A replacement of one string diagram by another that consists of a local modification, like in the picture (1.18), should be thought of as a *local move*. One benefit of the graphical string calculus is that such local moves can be used to successively simplify string diagrams. Another virtue is that the defining properties of morphisms can be readily accounted for. Indeed, associativity of composition is equivalent to locality of the move (1.18), whereby the morphisms $h \circ (g \circ f)$ and $(h \circ g) \circ f$ are both represented by the same string diagram, so that we are allowed to omit the brackets and write them both as $h \circ g \circ f$:

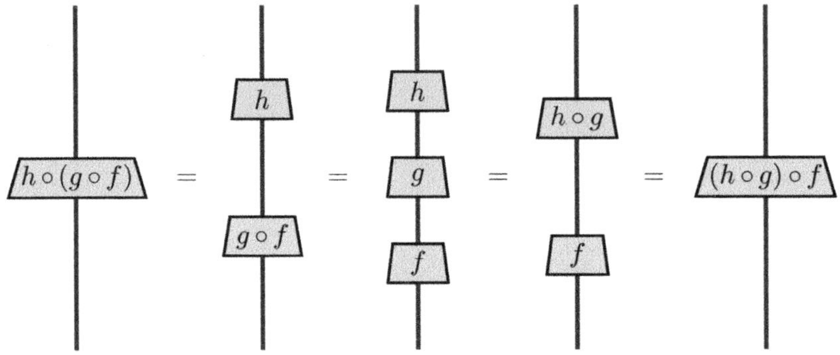

The significance of associativity is then that it allows one to replace in a unique manner any sequence of adjacent coupons by a single coupon. The properties of identity morphisms are accounted for in a similar manner, thereby also justifying the graphical convention (1.17).

Further below – in Graphical Descriptions 2.17, 2.60, and 2.94 – we will extend the graphical calculus so as to account for monoidal structures, braidings, and rigid structures. Introductions to the graphical calculus, including variants different from the one we are using here, abound in the literature. As a short selection we mention [**JS1**, **Kas**, **FeFFS**, **BaeS**, **Sel**, **TV1**].

In Definition 1.10 we have introduced the notion of isomorphism classes of objects of a category. Besides requiring that two objects are isomorphic, there is also another interesting way to regard them as the members of a class, namely by requiring instead that they just be connected by a finite sequence of morphisms, which do not have to be invertible and which are allowed to point in either direction. This weaker requirement is formalized as follows:

DEFINITION 1.19. Let $\mathcal{C}$ be a category. A *finite zig-zag* between objects X and X' in $\mathcal{C}$ is a sequence

$$X \xrightarrow{f_0} X_1 \xleftarrow{f_1} X_2 \xrightarrow{f_2} X_3 \xleftarrow{f_4} \dots \xleftarrow{f_n} X_n \xrightarrow{f_{n+1}} X'$$

of morphisms in $\mathcal{C}$. We say that two objects in $\mathcal{C}$ *can be connected* if there exists a finite zig-zag between them. This defines an equivalence relation on the objects of $\mathcal{C}$. The equivalence classes with respect to this relation are called the *path components* of $\mathcal{C}$. The set of path components of $\mathcal{C}$ is denoted by $\pi_0(\mathcal{C})$.

1.2. Groupoids and cobordisms

We are now going to present a few important examples of categories from geometry and topology.

EXAMPLE 1.20. Recall the cobordism category $\mathcal{C}ob_{1,0}$ introduced in Example 1.6. To determine the set $\pi_0(\mathcal{C}ob_{1,0})$ of path components of objects of $\mathcal{C}ob_{1,0}$ one can proceed as follows. For an object Y in $\mathcal{C}ob_{1,0}$, denote the zero-dimensional manifold obtained by reversing the orientation of every point in Y by $\overline{Y}$. Define two compact oriented zero-dimensional manifolds Y_1 and Y_2 to be equivalent if there exists a compact oriented one-manifold X with oriented boundary $\partial X = \overline{Y_1} \sqcup Y_2$.

For instance, with the pictorial conventions $\mathrm{pt}_+ \equiv \bullet_+$ and $\mathrm{pt}_- \equiv \bullet_-$ used in Example 1.1, the manifolds $Y_1 = \bullet_- \sqcup \bullet_+ \sqcup \bullet_+$ and $Y_2 = \bullet_+$ are equivalent, as witnessed by the one-manifold

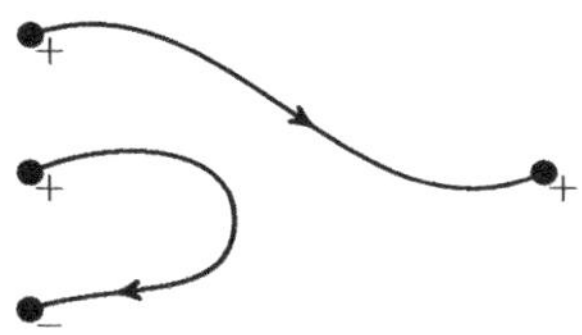

The so obtained set Ω_0^{SO} of equivalence classes of compact oriented zero-dimensional manifolds is, by construction, the set of path components:

$$\pi_0(\mathcal{C}ob_{1,0}) = \Omega_0^{\mathrm{SO}}.$$

EXERCISE 1.21. Show that Ω_0^{SO} has a natural structure of a group. (Indeed, Ω_0^{SO} is the free abelian group on one generator; a generator is the class of the positively oriented point pt_+.)
What group does one obtain when one replaces oriented manifolds by unoriented ones?

A specific type of category that will be particularly important to us is the following:

DEFINITION 1.22. A *groupoid* is a category in which all morphisms are isomorphisms.

EXERCISE 1.23. Given any category $\mathcal{C}$, keep all objects of $\mathcal{C}$, but restrict the morphisms by retaining only isomorphisms.
Show that this way one obtains a subcategory of $\mathcal{C}$.
We denote this subcategory by $\mathcal{C}^\times$. Show further that $\mathcal{C}^\times$ is a groupoid. We call $\mathcal{C}^\times$ the *maximal subgroupoid* of $\mathcal{C}$.
($\mathcal{C}^\times$ is an example of a *wide subcategory* of $\mathcal{C}$, i.e., of a subcategory that contains every object of $\mathcal{C}$.)

Thus in a groupoid $\mathcal{C}$ every morphism is invertible. It follows in particular that for a groupoid there is no difference between the set $\mathrm{Iso}(\mathcal{C})$ of isomorphism classes and the set $\pi_0(\mathcal{C})$ of path components.

EXAMPLES 1.24.

(1) A groupoid having a single object $*$ is completely characterized by the monoid $G := \mathrm{End}(*)$, which is a group. We write $*//G$ or, synonymously, $\mathcal{B}G$, for such a groupoid.
(2) An important type of groupoid is the *fundamental groupoid* of a topological space M, denoted by $\Pi_1(M)$. The objects of $\Pi_1(M)$ are the points of the space M. A morphism from $p \in M$ to $q \in M$ is a homotopy class of paths from p to q, where only homotopies relative to the boundary are allowed, i.e., homotopies that fix the starting and the endpoints of a path. In this groupoid the set $\mathrm{End}(p) = \pi_1(M,p)$ is the fundamental group for the base point $p \in M$. (An approach to topology through the perspective of the fundamental groupoid is presented in [**Bro**].)

(3) Consider a group G and a set X, together with an *action*

$$\rho: \quad G \times X \longrightarrow X\,, \qquad (g,x) \longmapsto g.x$$

of G on X, satisfying, by definition, $(gh).x = g.(h.x)$ for all $g, h \in G$ and all $x \in X$.

The *action groupoid* $X//G$ is the category whose objects are elements $x \in X$ and which has a morphism $x \to g.x$ for every pair $(g,x) \in G \times X$. (Note that we use the somewhat counterintuitive notation $X//G$ for a *left* action.)

(4) We call a set that is endowed with an action of a group G a *G-set*. The action ρ of G on a set X is called *transitive* if there is an element $x \in X$ such that every $y \in X$ can be written as $y = g.x$ for some $g \in G$; the action is called *free* if for every $x \in X$ the equality $g.x = x$ implies $g = e$. (As is readily checked, left multiplication endows any group G with the action of a transitive and free G-set.)

The *category G-Tor of G-torsors* is the category which has free and transitive G-sets as objects and G-equivariant maps as morphisms. That the category G-$\mathcal{T}or$ is a groupoid is seen as follows: Let $\varphi\colon X \to Y$ be a morphism of G-torsors. When working with right G-actions, the equivariance property $\varphi(x.g) = \varphi(x).g$ for all $x \in X$ and $g \in G$ implies that any point $y \in Y$ can be written in the form $\varphi(x_0).g = \varphi(x_0.g)$ for some fixed base point $x_0 \in X$ and some $g \in G$, and thus that φ is surjective. To show injectivity of φ, let $x, x' \in X$. Then $x' = x.g$ with a uniquely determined group element $g \in G$. Now $\varphi(x) = \varphi(x')$ implies $\varphi(x) = \varphi(x).g$. This, in turn, gives $g = e$ by freeness of the action on Y, and hence $x = xe = x'$.

EXERCISE 1.25. Show that any group G is a G-set when endowed with the adjoint action $g.x := gxg^{-1}$.

Is this action transitive?

Is it free?

EXERCISE 1.26. Verify that the action groupoid $X//G$ is indeed a groupoid.

Show further that the set of isomorphism classes of this groupoid is the *set-theoretic quotient*,

$$\pi_0(X//G) = X/G\,,$$

i.e., the set of orbits of the G-action.

(Accordingly one also refers to the action groupoid $X//G$ as a *weak quotient*. $X//G$ provides more information than the set X/G; in particular it contains the information about the stabilizers of the G-action.)

An important class of examples of topological field theories can be constructed with the help of ideas from *gauge theory*. A brief survey of the principles of gauge theory will be given in Section 2.1. For the moment, we will be content with reviewing the key mathematical notion that is needed for exploring the relevant gauge theories, namely principal fiber bundles with finite structure groups G.

The finiteness condition on the group is technically quite crucial: It essentially eliminates all problems that might arise from analytic subtleties. Specifically, we will have to deal with constructions that involve summations, and these are guaranteed to make sense only if G is finite. Despite the finiteness of the group G, the resulting theories will allow us to gain insight into some important aspects of field theories.

DEFINITION 1.27. Let G be a finite group and X be a smooth manifold. A *G-bundle*, or *G-principal bundle* on X is a smooth (not necessarily connected) manifold P together with a local diffeomorphism $P \xrightarrow{\pi} X$ and a smooth right action

$$P \times G \to P\,, \qquad (p, g) \mapsto p.g$$

such that $\pi(p.g) = \pi(p)$ for all $p \in P$ and $g \in G$, and such that the action is transitive and free on the fiber $\pi^{-1}(x)$ over each point $x \in X$.

In more detail, the latter means:

- Transitivity:
  For any two points $p, p' \in \pi^{-1}(x)$, there exists a $g \in G$ such that $p' = p.g$.
- Freeness:
  The equality $p = p.g$ implies that $g = e$ is the neutral element of G.

The group G is called the *structure group* of the principal bundle.

EXAMPLE 1.28. For any smooth manifold (and, more generally, for any topological space) X and any finite group G, the projection map $X \times G \longrightarrow X$ is a principal G-bundle if we equip $X \times G$ with the right G-action that is induced by the right regular action of G on itself. This principal G-bundle over X is called the *trivial principal G-bundle* over X.

EXAMPLE 1.29. As an example of a nontrivial principal bundle, a principal $\mathbb{Z}_3$-bundle over the circle S^1 is illustrated in the following picture:

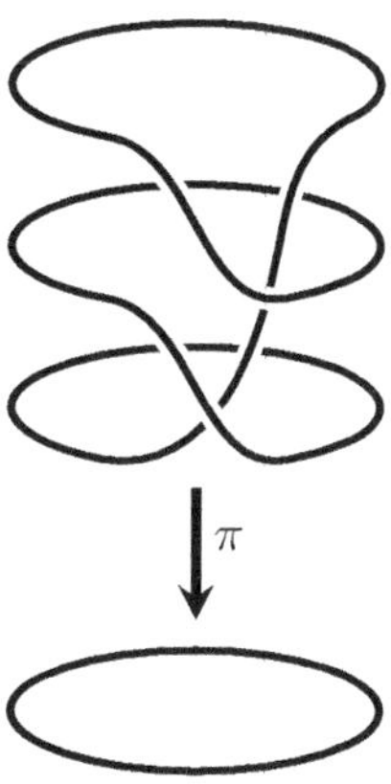

The local diffeomorphism π in the picture is generically given by vertical projection to the plane in which the circle S^1 is located; at those places where the lines are not horizontal, the pictorial representation is only symbolic.

DEFINITION 1.30. For any smooth manifold X and any finite group G, the *category $\mathcal{B}un_G(X)$ of G-bundles on X* is the category whose objects are G-bundles on X and whose morphisms

$$(P \xrightarrow{\pi} X) \longrightarrow (P' \xrightarrow{\pi'} X)$$

are smooth maps $\varphi\colon P \longrightarrow P'$ that commute with the G-actions, i.e., satisfy

$$\varphi(p.g) = \varphi(p).g$$

for $p \in P$ and $g \in G$, and cover the identity on X, i.e., $\pi' \circ \varphi = \pi$.

Expressed as a commutative diagram, the latter identity states commutativity of the triangle

$$\begin{array}{ccc} P & \xrightarrow{\varphi} & P' \\ & {\scriptstyle\pi}\searrow \quad \swarrow{\scriptstyle\pi'} & \\ & X & \end{array} .$$

EXERCISE 1.31. Set $X = S^1$, modeled as the unit circle in the complex plane, and $G = \mathbb{Z}_2$. Equip S^1 with the right $\mathbb{Z}_2$-action that is given by the involution $z \mapsto -z$.

Prove that the map $z \mapsto z^2$ from S^1 to itself is a principal $\mathbb{Z}_2$-bundle.

(As will follow easily from Proposition 1.73, the set of isomorphism classes of principal $\mathbb{Z}_2$-bundles over S^1 has in fact just two elements, namely the class of the trivial bundle from Example 1.28 and the class of the bundle considered here.)

EXAMPLE 1.32. The category of principal G-bundles over the topological space $*$ that consists of a single point is equivalent to the category G-$\mathcal{T}or$ of G-torsors and hence is a groupoid.

EXERCISE 1.33. Show that the category $\mathcal{B}un_G(X)$ of G-bundles over any manifold X is a groupoid.

EXERCISE 1.34. For G a finite group and X a manifold, let $\pi\colon P \longrightarrow X$ be a G-bundle over X.

(1) Show that for any path $\gamma\colon [0,1] \to X$ from x to y in X and for any fixed p in the fiber $P_x := \pi^{-1}(x)$ over x there is a unique *π-lift* of γ starting at p – that is, a unique path $\widetilde{\gamma}\colon [0,1] \longrightarrow P$ in P that descends to γ and starts at p, i.e., satisfies $\pi \circ \widetilde{\gamma} = \gamma$ and $\widetilde{\gamma}(0) = p$.

Hint: First establish that a principal bundle for a finite group is a covering of topological spaces, then use the *lifting theorems* for coverings. (See, e.g., Proposition 1.33 in [**Hat**].)

(2) Let $T^P_\gamma(p) := \widetilde{\gamma}(1)$ be the endpoint of $\widetilde{\gamma}$. Since this endpoint depends on the choice of p in the fiber P_x, varying p leads to a map $T^P_\gamma\colon P_x \longrightarrow P_y$ between the fibers over x and y. Show that T^P_γ is a morphism of G-torsors.

As we are working with a finite group, the map T^P_γ depends only on the homotopy class of γ relative to the endpoints (i.e., keeping the endpoints fixed); it is called the *(parallel) transport operator of P along γ*.

(3) Conclude that for a specified reference point $p \in P_x$ in the fiber over x the parallel transport along loops in X with base point $x \in X$ gives rise to a group homomorphism

$$\mathrm{hol}_p^P : \quad \pi_1(X, x) \longrightarrow G \tag{1.35}$$

from the fundamental group of (X, x) to G. This map is called the *holonomy morphism* of P relative to the reference point $p \in P$.
Show that if p is replaced by another point $p' = h.p \in P_x$ for some $h \in G$, then the holonomy homomorphism gets conjugated by h.
Show further that for P the trivial bundle, the holonomy morphism is the identity, $\mathrm{hol}_p^P(s) = e$, for every loop starting at x and every choice of reference point $p \in P$.

As a key ingredient for topological field theory we need the following notion, which generalizes Example 1.6 (and thereby supplies further examples of categories whose morphisms are not maps of any sort).

DEFINITION 1.36. Let d be any positive integer. The category $\mathcal{C}ob_{d,d-1}$ of d-dimensional oriented *cobordisms* (synonymously also called *bordisms*) is as follows:

(1) An object of $\mathcal{C}ob_{d,d-1}$ is a closed oriented smooth manifold of dimension $d-1$.
(2) Given a pair of objects $M, N \in \mathcal{C}ob_{d,d-1}$, a morphism $M \longrightarrow N$ is an equivalence class, with respect to the equivalence relation specified in (1.37) below, of cobordisms from M to N.
   Or, spelled out in more detail,
   - Let M and N be closed oriented $(d-1)$-dimensional smooth manifolds. A d-dimensional *cobordism* from M to N is an oriented, d-dimensional smooth manifold B with boundary, together with an orientation preserving diffeomorphism
   $$\phi_B : \quad \overline{M} \sqcup N \xrightarrow{\ \cong\ } \partial B \,.$$
   Here $\overline{M}$ is the same manifold as M but with opposite orientation.
   - Two cobordisms B and B' from M to N are defined to be equivalent, and thus to represent the same morphism in $\mathcal{C}ob_{d,d-1}$, if there is an orientation-preserving diffeomorphism $\phi\colon B \longrightarrow B'$ such that the diagram
   $$\begin{array}{ccc} B & \xrightarrow{\quad\phi\quad} & B' \\ & {}_{\phi_B}\nwarrow \quad \nearrow_{\phi'_B} & \\ & \overline{M} \sqcup N & \end{array} \tag{1.37}$$
   commutes.
   Here the maps $\overline{M} \sqcup N \to B$ and $\overline{M} \sqcup N \to B'$ are given by the composition of the diffeomorphisms $\phi_B\colon \overline{M} \sqcup N \to \partial B$ and $\phi'_B\colon \overline{M} \sqcup N \to \partial B'$ (which are the data of the cobordism) with the embeddings $\partial B \hookrightarrow B$ and $\partial B' \hookrightarrow B'$, respectively. By abuse of notation, we use the same symbols for these maps as for the underlying maps without the embeddings.
(3) The identity morphism for an object $M \in \mathcal{C}ob_{d,d-1}$ is represented by the product cobordism $B = M \times [0, 1]$.
   (Together with orientation-preserving diffeomorphisms from $\overline{M}$ to $M \times \{0\}$ and from M to $M \times \{1\}$ this is called the *cylinder* over M.)

(4) Composition of morphisms in $\mathcal{C}ob_{d,d-1}$ is given by *gluing* cobordisms: Given three objects $M, M', M'' \in \mathcal{C}ob_{d,d-1}$ and two cobordisms $B\colon M \longrightarrow M'$ and $B'\colon M' \longrightarrow M''$, the composition is defined to be the morphism that is represented by the manifold $B \sqcup_{M'} B'$.

(To get a smooth structure on this manifold, the gluing must actually be performed along a cylinder over M', often called a *collar*. Different choices of collars lead to different glued cobordisms. However, these are all diffeomorphic. We refer to Section 1.3 of [**Koc**] for the technical details.)

EXAMPLE 1.38. As described in Example 1.6, an object of $\mathcal{C}ob_{1,0}$ is a finite union of oriented points, and a morphism of $\mathcal{C}ob_{1,0}$ is an oriented one-dimensional manifold, possibly with boundary.

EXAMPLE 1.39. The objects of $\mathcal{C}ob_{2,1}$ are finite disjoint unions of oriented circles. There are six elementary morphisms from which every cobordism can be constructed by composition and disjoint union: the cylinder, the cap, the trinion or pair of pants and their inverses, and a pair of exchanging cylinders. For a presentation of the two-dimensional cobordism category in terms of generators and relations we refer to Section 3.2 (in particular to the pictures shown there and to Remark 3.6) and to [**Koc**].

REMARK 1.40. There are also various different variants of the cobordism category, which are of interest in other contexts. For instance, one may drop the requirement that the manifolds come equipped with an orientation and work with a *framing* instead. A framing of a d-dimensional manifold M is a d-tuple of nowhere vanishing and pointwise linearly independent tangent vector fields on M; the presence of a such a structure is equivalent to the existence of a trivialization of the tangent bundle of M.

One may also equip the geometrical category of cobordisms with further kinds of decorations, such as spin structures (see, e.g., [**MooS1**, Ex. 1]) or principal fiber bundles. Another possibility is to take the objects to be manifolds with boundary, and the morphisms to be manifolds with corners; this will be discussed in Section 5.2.

1.3. Functors and natural transformations

Analogously as in set theory, in which one not only considers individual sets, but also maps between them, it is of much interest to not just study individual categories, but also maps between categories. Since a category has two layers of data (objects and morphisms), such a structure is two-layered as well, involving mappings between objects as well as between morphisms. This is formalized in the following notion.

DEFINITION 1.41. Let $\mathcal{C}$ and $\mathcal{C}'$ be categories. A *functor* $F\colon \mathcal{C} \longrightarrow \mathcal{C}'$ consists of two maps

$$F: \quad \mathrm{Obj}(\mathcal{C}) \longrightarrow \mathrm{Obj}(\mathcal{C}') \qquad \text{and} \qquad F: \quad \mathrm{Hom}(\mathcal{C}) \longrightarrow \mathrm{Hom}(\mathcal{C}')$$

that are compatible with identity morphisms and composition of morphisms, in the sense that they obey the following conditions:

(1) For every object $V \in \mathrm{Obj}(\mathcal{C})$ one has $F(\mathrm{id}_V) = \mathrm{id}_{F(V)}$.

(2) For every morphism $f \in \mathrm{Hom}(\mathcal{C})$, one has
$s(F(f)) = F(s(f))$ and $t(F(f)) = F(t(f))$.

(3) For any pair f, g of composable morphisms one has

$$F(g \circ f) = F(g) \circ F(f) .$$

Given two functors

$$F: \quad \mathcal{C} \longrightarrow \mathcal{C}' \qquad \text{and} \qquad G: \quad \mathcal{C}' \longrightarrow \mathcal{C}'',$$

their composition, or concatenation, $G \circ F : \mathcal{C} \longrightarrow \mathcal{C}''$ is the functor that is obtained by composition of maps.

Since composition of maps is associative, the concatenation of functors is associative: one has

$$H \circ (G \circ F) = (H \circ G) \circ F$$

for any triple of functors $F: \mathcal{C}_1 \longrightarrow \mathcal{C}_2$, $G: \mathcal{C}_2 \longrightarrow \mathcal{C}_3$, and $H: \mathcal{C}_3 \longrightarrow \mathcal{C}_4$.

A functor $F: \mathcal{C} \longrightarrow \mathcal{C}$ is called an *endofunctor* of $\mathcal{C}$. For a functor between categories that are enriched over $\mathcal{V}ect(\mathbb{K})$, it is natural to require that the functor has to be compatible with this structure, i.e., is $\mathbb{K}$-linear on morphism spaces; such a functor is called a $\mathbb{K}$-*linear* functor.

Recall the notion of *opposite* category; see Example 1.5(1). For given categories $\mathcal{C}$ and $\mathcal{D}$, a functor $F: \mathcal{C}^{\mathrm{opp}} \longrightarrow \mathcal{D}$ is referred to as a *contravariant* functor from $\mathcal{C}$ to $\mathcal{D}$. Note that a functor $\mathcal{C}^{\mathrm{opp}} \longrightarrow \mathcal{D}$ can equivalently be described as a functor $\mathcal{C} \longrightarrow \mathcal{D}^{\mathrm{opp}}$. In short, a contravariant functor reverses the arrows.

Here are a few examples of functors, some of which we have actually already encountered above:

EXAMPLES 1.42.

(1) For any category $\mathcal{C}$, the *identity functor* $\mathrm{id}_{\mathcal{C}}: \mathcal{C} \longrightarrow \mathcal{C}$, which acts as the identity map both on objects and morphisms, is an endofunctor.

(2) Mapping any group G to the underlying set (i.e., forgetting all the group structure) and any group homomorphism to the underlying map of sets gives a functor from the category of groups to the category of sets, called the *forgetful functor.*

Likewise, for any category $\mathcal{C}$ whose objects and morphisms are sets, and maps between sets, respectively, that are endowed with any kind of additional structure, there is an analogous forgetful functor from $\mathcal{C}$ to the category of sets.

(3) For any category $\mathcal{C}$, $\mathrm{Hom}: \mathcal{C}^{\mathrm{opp}} \times \mathcal{C} \longrightarrow \mathcal{S}et$ is a functor that sends an ordered pair (X, Y) of objects to the morphism set $\mathrm{Hom}_{\mathcal{C}}(X, Y)$. A morphism $(X, Y) \longrightarrow (X', Y')$ in the category $\mathcal{C}^{\mathrm{opp}} \times \mathcal{C}$ is a pair (f, g) of morphisms $f: X' \longrightarrow X$ and $g: Y \longrightarrow Y'$ in $\mathcal{C}$. The functor Hom sends it to the map

$$\begin{aligned} \mathrm{Hom}(f, g): \quad \mathrm{Hom}_{\mathcal{C}}(X, Y) &\longrightarrow \mathrm{Hom}_{\mathcal{C}}(X', Y') , \\ h &\longmapsto g \circ h \circ f . \end{aligned}$$

(4) Given a field $\mathbb{K}$ and a group G, a functor $*/\!/G \longrightarrow \mathcal{V}ect(\mathbb{K})$ amounts to a $\mathbb{K}$-linear representation of G.

Similarly, given an associative $\mathbb{K}$-algebra A, a $\mathbb{K}$-linear functor

$$*/\!/A \longrightarrow \mathcal{V}ect(\mathbb{K})$$

amounts to an A-module. Explicitly, this is a $\mathbb{K}$-vector space M together with a morphism $\rho_M: A \longrightarrow \mathrm{End}_{\mathbb{K}}(M)$ of $\mathbb{K}$-algebras.

(5) Associating to a $\mathbb{K}$-vector space V its dual space V^* provides a functor

$$\begin{aligned} \mathcal{V}ect(\mathbb{K}) &\longrightarrow \mathcal{V}ect(\mathbb{K})^{\mathrm{opp}}, \\ V &\longmapsto V^* \end{aligned}$$

from the category $\mathcal{V}ect(\mathbb{K})$ to its opposite category.
Taking the double dual provides an endofunctor

$$\begin{aligned} (-)^{**}: \quad \mathcal{V}ect(\mathbb{K}) &\longrightarrow \mathcal{V}ect(\mathbb{K}), \\ V &\longmapsto V^{**} \end{aligned} \tag{1.43}$$

of $\mathcal{V}ect(\mathbb{K})$.

(6) Let $\phi\colon R \longrightarrow S$ be a morphism of rings. Then to any given S-module (M, ρ) we can associate an R-module ϕ^*M that is defined as the abelian group M together with the action of R given by the composite

$$R \times M \xrightarrow{\phi \times \mathrm{id}_M} S \times M \xrightarrow{\rho} M .$$

The map underlying a morphism $\varphi\colon M \longrightarrow N$ of S-modules also provides a morphism $\phi^*M \longrightarrow \phi^*N$ of R-modules. We thus have a functor

$$\phi^*: \quad S\text{-mod} \longrightarrow R\text{-mod}.$$

By a slight abuse of language, ϕ^* is called the *restriction functor* from S-mod to R-mod, or also the *restriction of scalars*.

EXERCISE 1.44. Given a functor $F: \mathcal{C} \longrightarrow \mathcal{D}$ and an object $d \in \mathcal{D}$, the *slice category over d*, denoted by F/d, is defined as follows. An object of F/d is a pair (c, f) consisting of an object $c \in \mathcal{C}$ and a morphism $f: F(c) \longrightarrow d$ in $\mathcal{D}$. A morphism $(c, f) \longrightarrow (c', f')$ between two such pairs is a morphism $g\colon c \longrightarrow c'$ in $\mathcal{C}$ such that $f' \circ F(g) = f$. In the special case $F = \mathrm{id}_{\mathcal{C}}$ we write $\mathcal{C}/c := \mathrm{id}_{\mathcal{C}}/c$ for $c \in \mathcal{C}$.

(1) For any functor $F: \Gamma \longrightarrow \Omega$ between groupoids and any object $y \in \Omega$, describe the slice category F/y. More precisely, what are the isomorphism classes and automorphism groups of F/y?
(2) An isomorphism class of a monomorphism in a category $\mathcal{C}$ with target c, seen as an object in $\mathcal{C}/c$, is called a *subobject* of c. Describe the subobjects of all objects in the category of finite-dimensional vector spaces. What are the finite-dimensional vector spaces with exactly one or exactly two subobjects?

In the following exercise we show that a morphism $X \xrightarrow{f} Y$ of smooth manifolds gives rise to a functor $f^*: \mathcal{B}un_G(Y) \longrightarrow \mathcal{B}un_G(X)$ that relates the groupoids of G-bundles which we studied in Exercise 1.33.

EXERCISE 1.45.

(1) Let $X \xrightarrow{f} Y$ be a morphism of smooth manifolds and $P \xrightarrow{\pi} Y$ a G-bundle on Y. Show that the *pullback*, i.e., the set

$$X \times_Y P := \{ (x, p) \in X \times P \mid f(x) = \pi(p) \}$$

endowed with the subspace topology, i.e., with the coarsest topology such that the embedding is continuous, is a smooth submanifold of $X \times P$.
(The pullback depends on the maps f and π, which are omitted in the notation. Also, it is important that here $P \xrightarrow{\pi} Y$ is a bundle. In general, the pullback of manifolds is not a manifold, as the following example shows: Let $Y = \mathbb{R}$ be

a one-dimensional manifold, $X=\{0\}$ a zero-dimensional manifold, and f the natural embedding. Take $P=\mathbb{R}^2$ and $\pi(x,y)=xy$. Then $X\times_Y P$ is the subset $\{(x,y)\,|\,xy=0\}$ in $\mathbb{R}^2$, which is not a manifold.)

(2) Show that the pullback f^*P has the following universal property: Let Z be a smooth manifold and α_X and α_P be morphisms of smooth manifolds such that the square

$$\begin{array}{ccc} Z & \xrightarrow{\alpha_P} & P \\ {\scriptstyle\alpha_X}\downarrow & & \downarrow{\scriptstyle\pi} \\ X & \xrightarrow{f} & Y \end{array}$$

commutes. Then there exists a unique map α of smooth manifolds such that also the two triangles in the diagram

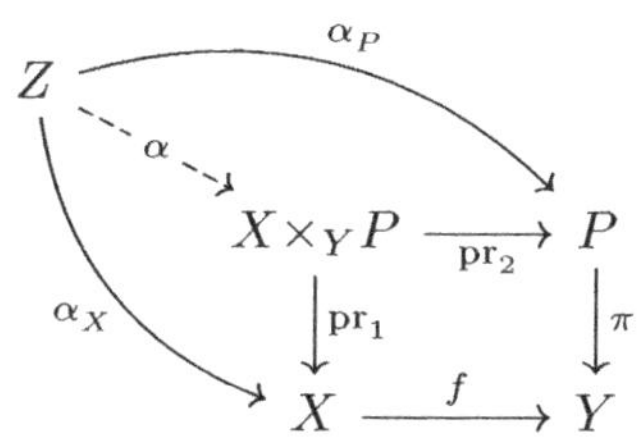

commute. (Here pr_1 and pr_2 denote the projection on the first and second factor of the product manifold $X\times P$, respectively.)

(3) Show further that the G-action on P given by definition induces a G-action on the manifold $X\times_Y P$, whereby $X\times_Y P$ forms a G-bundle on X that makes the square

$$\begin{array}{ccc} X\times_Y P & \xrightarrow{\mathrm{pr}_2} & P \\ {\scriptstyle\mathrm{pr}_1}\downarrow & & \downarrow{\scriptstyle\pi} \\ X & \xrightarrow{f} & Y \end{array}$$

commute. The bundle $X\times_Y P$ is denoted by f^*P.

(4) Conclude that, given a morphism $X\xrightarrow{f}Y$ of smooth manifolds, we have in this way constructed a functor

$$f^*:\quad \mathcal{B}un_G(Y)\longrightarrow \mathcal{B}un_G(X)\,.$$

Hint: For a morphism $\varphi\colon P\longrightarrow P'$ of G-bundles on Y, use the universal property of the pullback in the diagram

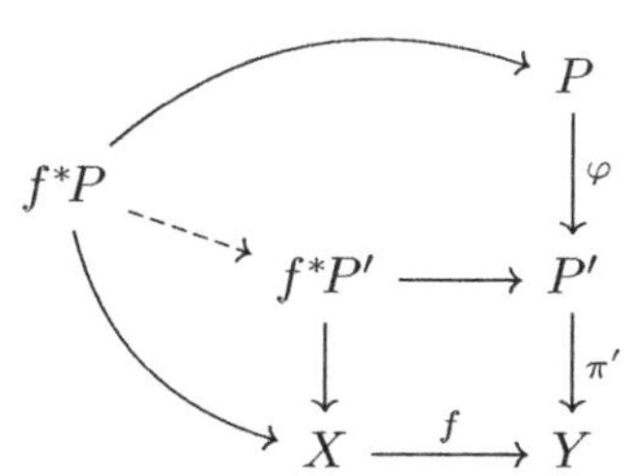

Group morphisms give rise to functors between groupoids of bundles as well:

EXERCISE 1.46. Let $\varphi\colon H \to G$ be a morphism of finite groups. Then for a H-bundle $\pi\colon P \to X$ over X we define

$$P \times_\varphi G := P \times G/\!\sim,$$

where $\sim$ is the relation $(p.h, g) \sim (p, \varphi(h)g)$. ($P \times_\varphi G$ is taken to be endowed with the quotient topology, i.e., with the finest topology such that the surjection is continuous.)

Show that π induces a map $P \times_\varphi G \to X$ which is continuous and is even a G-bundle. We call this bundle the *extension of P along φ* and denote it by $\mathrm{Ind}_H^G(P) \equiv \mathrm{Ind}_{H;\varphi}^G(P)$.

Observe further that the assignment $P \mapsto \mathrm{Ind}_H^G(P)$ extends to a functor

$$\mathrm{Ind}_H^G : \quad \mathcal{B}un_H(X) \longrightarrow \mathcal{B}un_G(X)\,.$$

As seen in Example 1.42(4), a representation of a group G is the same as a functor $*/\!/G \to \mathcal{V}ect(\mathbb{K})$. Since the representations of a group, together with intertwiners between representations as morphisms, form a category, this suggests that there should be a notion of morphism between functors and thereby a category of functors. This idea is formalized by the following concept:

DEFINITION 1.47. Let $F, G\colon \mathcal{C} \to \mathcal{C}'$ be functors between the same pair of categories. A *natural transformation*

$$\eta : \quad F \longrightarrow G$$

from F to G is a family of morphisms

$$\eta_V : \quad F(V) \longrightarrow G(V)$$

in $\mathcal{C}'$, indexed by $V \in \mathrm{Obj}(\mathcal{C})$ and called the *components* of η, such that for any morphism $f\colon V \to W$ in $\mathcal{C}$ the diagram

$$\begin{array}{ccc} F(V) & \xrightarrow{\ \eta_V\ } & G(V) \\ {\scriptstyle F(f)}\big\downarrow & & \big\downarrow{\scriptstyle G(f)} \\ F(W) & \xrightarrow{\ \eta_W\ } & G(W) \end{array} \tag{1.48}$$

in $\mathcal{C}'$ commutes.

If for every object $V \in \mathrm{Obj}(\mathcal{C})$ the morphism η_V is an isomorphism, then the natural transformation $\eta\colon F \to G$ is called a *natural isomorphism*.

Natural transformations can be composed in two different ways:

DEFINITION 1.49. For functors $F, F', F''\colon \mathcal{C} \to \mathcal{D}$, the *vertical composition*

$$\eta' \circ \eta : \quad F \to F''$$

of two natural transformations $\eta\colon F \to F'$ and $\eta'\colon F' \to F''$ is the natural transformation that is obtained by composing their component morphisms, i.e.

$$F(X) \xrightarrow{(\eta' \circ \eta)_X} F''(X) \ := \ F(X) \xrightarrow{\eta_X} F'(X) \xrightarrow{\eta'_X} F''(X)$$

for any object X in $\mathcal{C}$.

For functors $F, F'\colon \mathcal{C} \to \mathcal{C}'$, and $G, G'\colon \mathcal{C}' \to \mathcal{C}''$, the horizontal composition

$$\eta \star \varphi : \quad G \circ F \longrightarrow G' \circ F'$$

of two natural transformations $\varphi\colon F \longrightarrow F'$ and $\eta\colon G \longrightarrow G'$ is the natural transformation that in terms of components is given by

$$(\eta \star \varphi)_X := \eta_{F'X} \circ G(\varphi_X) = G'(\varphi_X) \circ \eta_{FX}$$

for any object X in $\mathcal{C}$.

EXERCISE 1.50. Verify that the so defined vertical and horizontal composites $\eta' \circ \eta$ and $\eta \star \varphi$ are indeed natural transformations from F to F'' and from $G \circ F$ to $G' \circ F'$, respectively.

In particular, prove that the equality $(\eta \star \varphi)_X = G'(\varphi_X) \circ \eta_{FX}$ holds for every object X.

We visualize a natural transformation $\eta\colon F \longrightarrow G$ by the diagram

(1.51)

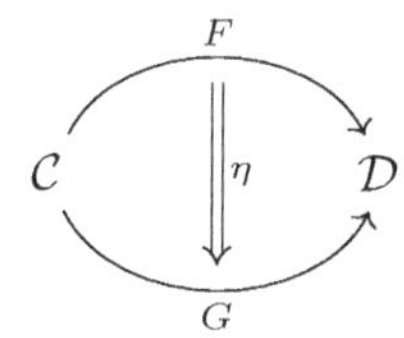

According to this description, it is appropriate to think of η as being assigned to the 2-gon formed by the two arrows; a natural transformation is therefore sometimes referred to as a *filler* of the diagram (1.51).

In terms of such diagrams, the vertical composition of natural transformations η and η' is expressed as

(1.52)

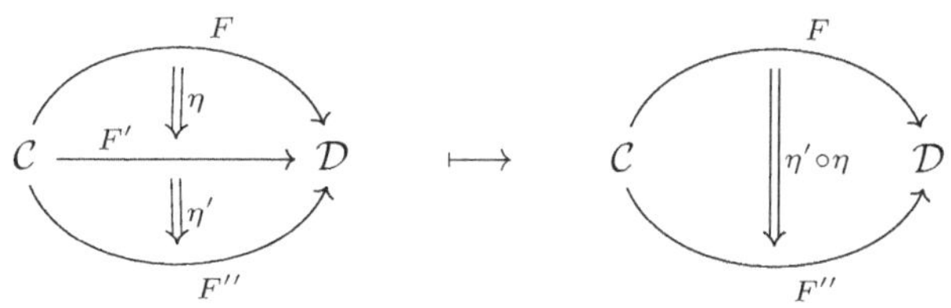

and the horizontal composition as

(1.53)

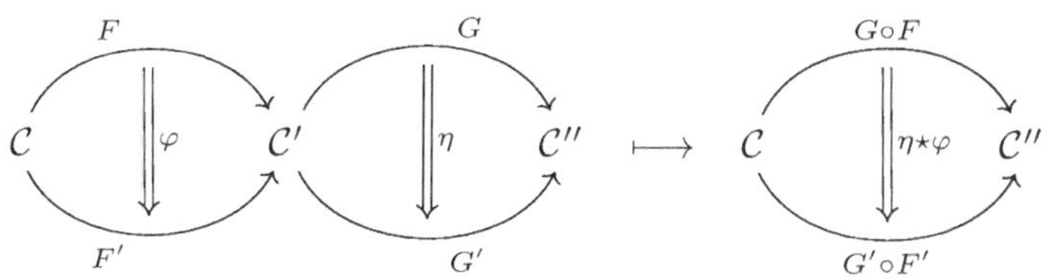

EXAMPLES 1.54 ([**AdHS**, 6.2]).

(1) Consider the following "squaring functor" Sq from the category of groups to the category of sets: $\mathrm{Sq}(G) := G \times G$ for any group G and $\mathrm{Sq}(f) := f \times f$ for any group homomorphism $f\colon G \longrightarrow H$. For a group G, the multiplication is a map $\mu_G\colon G \times G \longrightarrow G$. The family $\mu = (\mu_G)$ is a natural transformation from Sq to the forgetful functor. Indeed, the naturality condition (1.48) just amounts to $f(\mu_G(g, g')) = \mu_H(f(g), f(g'))$ for all group homomorphisms $f\colon G \longrightarrow H$ and all $g, g' \in G$.

(2) The same construction works when the category of groups is replaced by the category of any type of algebras.

(3) For a vector space V define the linear map η_V from V to its double dual V^{**} by $(\eta_V(v))(f) := f(v)$ for every $v \in V$ and $f \in V^*$. The family (η_V) is a natural transformation from the identity functor of $\mathcal{V}ect$ to the double dual functor (1.43).

(4) In algebraic topology, for any $n \in \mathbb{Z}_{\geq 1}$ there are the functors π_n and H_n from the category of topological spaces to the category of groups, which map a topological space X to its nth homotopy group and nth homology group, respectively. Assigning to every topological space X the *Hurewicz homomorphism* from $\pi_n(X)$ to $H_n(X)$ gives a natural transformation $\pi_n \longrightarrow H_n$ between these two functors.

(5) For any category $\mathcal{C}$ and any morphism $f \in \mathrm{Hom}_{\mathcal{C}}(c, c')$, pre-composition with f defines a natural transformation from the Hom functor $\mathrm{Hom}_{\mathcal{C}}(c', -)$ to the Hom functor $\mathrm{Hom}_{\mathcal{C}}(c, -)$. Post-composition with f defines a natural transformation from $\mathrm{Hom}_{\mathcal{C}}(-, c)$ to $\mathrm{Hom}_{\mathcal{C}}(-, c')$.

EXAMPLE 1.55. Let $F\colon \mathcal{C} \longrightarrow \mathcal{S}et$ be a functor from a category $\mathcal{C}$ to the category of sets and let X be an object of $\mathcal{C}$. Denote by $\mathrm{Nat}(\mathrm{Hom}_{\mathcal{C}}(X, -), F)$ the set of natural transformations from the functor $\mathrm{Hom}_{\mathcal{C}}(X, -)\colon \mathcal{C} \longrightarrow \mathcal{S}et$ to F. Then the map that sends a natural transformation $\eta \in \mathrm{Nat}(\mathrm{Hom}_{\mathcal{C}}(X, -), F)$ to the value $\eta_X(\mathrm{id}_X) \in F(X)$ of the component map $\eta_X\colon \mathrm{Hom}_{\mathcal{C}}(X, X) \longrightarrow F(X)$ yields a bijection

$$\mathrm{Nat}(\mathrm{Hom}_{\mathcal{C}}(X, -), F) \xrightarrow{\cong} F(X)\,.$$

The inverse map $F(X) \longrightarrow \mathrm{Nat}(\mathrm{Hom}_{\mathcal{C}}(X, -), F)$ maps an element x of the set $F(X)$ to the natural transformation from $\mathrm{Hom}_{\mathcal{C}}(X, -)$ to F whose component at $Y \in \mathcal{C}$ is the map $\mathrm{Hom}_{\mathcal{C}}(X, Y) \longrightarrow F(Y)$ that sends $f\colon X \longrightarrow Y$ to $(F(f))(x)$.

This is one of the many variants of what in category theory is referred to as the *Yoneda lemma*. For further background we refer to Section 2.2 of [**Ri2**].

Natural transformations also play an important role in the context of bundles. This is illustrated in the following example.

EXAMPLE 1.56. Let G be a finite group. As seen in Exercise 1.45(4), for any map $f\colon M \longrightarrow N$ of manifolds there is a pullback functor $f^*\colon \mathcal{B}un_G(N) \longrightarrow \mathcal{B}un_G(M)$. Let $f_0, f_1\colon M \longrightarrow N$ be two such maps, and let α be a *homotopy* from f_0 to f_1, that is, a map $\alpha\colon M \times [0.1] \longrightarrow N$ satisfying

$$\alpha(m, 0) = f_0(m) \qquad \text{and} \qquad \alpha(m, 1) = f_1(m)$$

for each $m \in M$. For any point $m \in M$, consider the path s_m in N from $s_m(0) = f_0(m)$ to $s_m(1) = f_1(m)$ that is given by $s_m(t) := \alpha(m, t)$ for $t \in [0, 1]$. For P_G a G-bundle on N, *parallel transport* along the path s_m induces a map $(P_G)_{f_0(m)} \xrightarrow{\cong} (P_G)_{f_1(m)}$ of fibers of the bundle. The corresponding maps on the fibers $(f_0^* P_G)_n$ and $(f_1^* P_G)_n$ of the pullbacks combine into an isomorphism

$$\alpha_{P_G}: \quad f_1^* P_G \xrightarrow{\cong} f_0^* P_G$$

of G-bundles. This is natural in bundle morphisms. Hence the homotopy α provides a natural transformation

$$\alpha^*: \quad f_0^* \longrightarrow f_1^*$$

between the two pullback functors.

We can now return to our original motivation for introducing the concept of a natural transformation: A functor $F\colon *//G \longrightarrow \mathcal{V}ect(\mathbb{K})$ evaluated on the single object $*$ gives a vector space $V := F(*)$ and a morphism $G \longrightarrow \mathrm{End}(V)$ that endows V with the structure of a G-representation. A natural transformation between functors F and F' from $*//G$ to $\mathcal{V}ect(\mathbb{K})$ has a single component corresponding to the single object $*$, which constitutes a linear map $\eta_*\colon F(*) \longrightarrow F'(*)$. The commuting diagram (1.48) then implies that η_* is an intertwiner between the G-representations $F(*)$ and $F'(*)$.

EXERCISE 1.57. (1) Let $\mathcal{C}$ be a *small* category, i.e., such that $\mathrm{Obj}(\mathcal{C})$ is a set, and let $\mathcal{D}$ be any category. Show that the functors $\mathcal{C} \longrightarrow \mathcal{D}$ together with natural transformations form a category, to be denoted as $[\mathcal{C},\mathcal{D}]$ and called the *category of functors from $\mathcal{C}$ to $\mathcal{D}$*.

(The condition that $\mathcal{C}$ is small ensures that the morphism sets of the functor category are indeed sets, rather than just classes.)

(2) Let G be a finite group and $\mathbb{K}$ be a field. Show that the functor category $[*//G, \mathcal{V}ect(\mathbb{K})]$ is just the category of G-representations.

In view of these observations we will refer to a functor $\mathcal{C} \longrightarrow \mathcal{V}ect(\mathbb{K})$ also as a $\mathbb{K}$-linear *representation* of the category $\mathcal{C}$.

EXAMPLE 1.58 (*Toward topological field theory*). Of central interest to us will be a $\mathbb{K}$-linear representation of the cobordism category $\mathcal{C}ob_{d,d-1}$. We introduce it in the following first attempt at defining the notion of a *topological field theory* in dimension d:

Tentative Definition. A *d-dimensional topological field theory* is a representation of a d-dimensional cobordism category, i.e., it is a functor

$$Z:\quad \mathcal{C}ob_{d,d-1} \longrightarrow \mathcal{V}ect(\mathbb{K})\,. \tag{1.59}$$

This means in particular that we assign vector spaces to $(d-1)$-dimensional manifolds and linear maps to d-dimensional cobordisms. To promote this first ansatz to a full definition, it will be necessary to identify further structure with which the cobordism category as well as the category of vector spaces can be endowed. The topological field theory functor is then required to be compatible with those structures. We will include the extra structures in question in several steps; see (2.88) for the next step, and Definition 2.100 for the final version of the notion.

As pointed out at the beginning of the invitation chapter, in the terminology of quantum field theory the two layers of structure we are dealing with have the following interpretation: the vector spaces that are assigned to $(d-1)$-manifolds are spaces of field configurations which play the role of initial conditions, while the linear maps assigned to cobordisms are transition amplitudes. To obtain a concrete model of a quantum field theory that fits into this picture is in fact far from trivial. As we will explain in Section 2.1, a suitable framework for this specific class of models is provided by a particular kind of gauge theories, for which the field configurations form a groupoid rather than a set.

We continue to develop our general categorical tools. The notion of natural isomorphism allows us to formulate what it means that two categories are essentially the same:

DEFINITION 1.60. A functor

$$F: \quad \mathcal{C} \longrightarrow \mathcal{D}$$

is called an *equivalence of categories* if there exist a functor

$$G: \quad \mathcal{D} \longrightarrow \mathcal{C}$$

and natural isomorphisms

$$\begin{aligned} \eta: &\quad \mathrm{id}_{\mathcal{D}} \longrightarrow F \circ G \\ \text{and} \qquad \theta: &\quad G \circ F \longrightarrow \mathrm{id}_{\mathcal{C}}. \end{aligned}$$

REMARK 1.61. Invoking the general principle, formulated in Remark 1.12, that isomorphisms of objects are more significant than equalities, we learn that usually it does not make sense to require two functors to be equal, but we should rather require them to be isomorphic. Accordingly, the natural relation between categories that are "essentially the same" is equivalence of categories. As a consequence, "reasonable" properties of a category should be shared by all categories that are equivalent to it. This is called the *principle of equivalence.*

For example, the number of isomorphism classes or path components is invariant under equivalence of categories and is therefore a reasonable quantity that we can extract from a category. In contrast, the number of objects of a category is not invariant under equivalence and is thus not a reasonable quantity.

The following result provides a convenient criterion for detecting equivalences of categories:

LEMMA 1.62. A functor $F\colon \mathcal{C} \longrightarrow \mathcal{D}$ is an equivalence of categories if and only if F has both of the following properties:

(1) The functor F is *essentially surjective*, i.e., for any object W of $\mathcal{D}$ there exists an object V of $\mathcal{C}$ such that $F(V) \cong W$ in $\mathcal{D}$.
(2) The functor F is *fully faithful*, i.e., for any pair V, V' of objects in $\mathcal{C}$ the map

$$F: \quad \mathrm{Hom}_{\mathcal{C}}(V, V') \longrightarrow \mathrm{Hom}_{\mathcal{D}}(F(V), F(V'))$$

on morphism sets is bijective.

PROOF. A complete proof can be found, e.g., in [**Kas**, p. 278]. Let us just mention here that when constructing a quasi-inverse to a fully faithful and essentially surjective functor one has to use the axiom of choice. □

The notion of equivalence allows us to introduce further concepts for categories: A category is called *essentially small* if it is equivalent to a small category. A category is called *skeletal* if any two of its objects that are isomorphic are in fact equal. A *skeleton* of a category $\mathcal{C}$ is a skeletal subcategory of $\mathcal{C}$ such that the inclusion functor is an equivalence of categories. By picking one object in each isomorphism class we see that any category has a skeleton. (Here it is assumed that the axiom of choice holds.)

EXERCISE 1.63.

(1) Show that the family of natural maps $\iota_V : V \longrightarrow V^{**}$, for V any $\mathbb{K}$-vector space, defined by $\iota_V(v)(\varphi) := \varphi(v)$ for $v \in V$ and $\varphi \in V^*$ forms a natural transformation

$$\iota : \quad \mathrm{id}_{\mathcal{V}ect(\mathbb{K})} \longrightarrow (-)^{**}$$

of endofunctors on $\mathcal{V}ect(\mathbb{K})$.

(2) Restrict these functors to the full subcategory $\mathcal{V}ect_{\text{f.d.}}(\mathbb{K})$ of finite-dimensional $\mathbb{K}$-vector spaces.

Show that the natural transformation that is induced on the resulting restrictions is a natural isomorphism.

For later purposes it is crucial that we can replace the groupoid of G-bundles over a manifold M by an equivalent groupoid that is built from very basic topological data of M. Such a description will be derived in Example 1.74 for G-bundles on the circle S^1. This will be used to obtain, in Equation (5.32), a compact algebraic expression for the category that is assigned to S^1 by a three-dimensional Dijkgraaf-Witten field theory.

The following considerations prepare the statement and proof of this result.

EXERCISE 1.64. Verify that for a disjoint union $Y_1 \sqcup Y_2$ of smooth manifolds the inclusions of Y_1 and Y_2 into the disjoint union induce a functor

$$\mathcal{B}un_G(Y_1 \sqcup Y_2) \longrightarrow \mathcal{B}un_G(Y_1) \times \mathcal{B}un_G(Y_2) \ .$$

Prove that this functor is an equivalence.

EXERCISE 1.65. Let X be a connected manifold or, more generally, a path-connected topological space. Any choice of a base point $x_0 \in X$ gives a functor

$$\iota_{x_0} : \quad *//\pi_1(X, x_0) \longrightarrow \Pi_1(X)$$

from the groupoid with one object $*$ and automorphism group $\pi_1(X, x_0)$ to the fundamental groupoid $\Pi_1(X)$ introduced in Example 1.24(2), as follows: The functor ι_{x_0} sends the single object $*$ to the base point $x_0 \in X$ and sends a closed path to the corresponding endomorphism of x_0 in $\Pi_1(X)$.

Show that the functor ι_{x_0} is an equivalence of categories.

1.4. Groupoids of bundles

We now use the categorical tools just developed to describe groupoids whose objects are bundles. The results of this section will play a crucial role in the study of Dijkgraaf-Witten topological field theories, in particular in Sections 5.4 and 5.5. Some readers might mainly take notice of Proposition 1.73 and Example 1.74 at first.

In Exercise 1.33 we have seen that for any manifold X and any finite group G the category $\mathcal{B}un_G(X)$ of G-bundles on X is a groupoid. We have also seen, in Exercise 1.45, that any morphism $f : X \longrightarrow Y$ of manifolds gives rise to a pullback functor $f^* : \mathcal{B}un_G(Y) \longrightarrow \mathcal{B}un_G(X)$. Finally, as explained in Example 1.56, given a homotopy α between two maps $f_1, f_2 : X \longrightarrow Y$ we get a natural isomorphism $\alpha^* : f_1^* \longrightarrow f_2^*$ between the pullback functors. G-bundles are thus well adapted to the category $\mathcal{M}an$ of manifolds. (In fact, they form a *stack* on $\mathcal{M}an$.) This turns them into a convenient tool for constructing examples of topological field theories. In the present section we aim at obtaining a concrete algebraic understanding of these groupoids of bundles. This will eventually allow us to extract explicit algebraic structures from topological field theories constructed from bundles.

Let X be a manifold and G be a finite group. Recall from Exercise 1.34(3) that for any G-bundle P over X the fiber over any point $x \in X$ is a G-torsor, and that for any path $\gamma\colon [0,1] \to X$, parallel transport gives a morphism $\mathrm{hol}_\gamma\colon P_{\gamma(0)} \to P_{\gamma(1)}$ of G-torsors which depends only on the homotopy class of γ. Now recall that homotopy classes of paths are the morphisms in the fundamental groupoid $\Pi_1(X)$ of X that was introduced in Exercise 1.24(2). Accordingly we get for any G-bundle P over X a transport functor

$$T^P: \quad \Pi_1(X) \longrightarrow G\text{-}\mathcal{Tor}$$

from the fundamental groupoid of X to the groupoid of G-torsors. This yields a functor

$$T: \quad \mathcal{Bun}_G(X) \longrightarrow [\Pi_1(X), G\text{-}\mathcal{Tor}]$$

between groupoids. This assignment is completely geometrical in the sense that it does not involve any choices.

We will now show that the functor T is an equivalence. To arrive at this result, we first notice that by invoking Exercise 1.64 we can restrict ourselves to the situation that the base manifold X is connected. From now on we assume that this is the case. A connected manifold is also path-connected. Accordingly, when X is connected, we would rather prefer to work with the fundamental group $\pi_1(X, x_0)$ for some chosen base point $x_0 \in X$, for which frequently an explicit presentation in terms of generators and relations is available. The fundamental groups for different base points are isomorphic, but not canonically isomorphic. Indeed, as seen in Exercise 1.65, each choice of a base point $x_0 \in X$ gives an equivalence

$$\iota_{x_0}: \quad *//\pi_1(X, x_0) \longrightarrow \Pi_1(X)$$

of categories between the groupoid that has a single object with automorphism group $\pi_1(X, x_0)$ and the fundamental groupoid of X.

DEFINITION 1.66. For X a connected manifold and a choice of a point $x_0 \in X$, the category $\mathcal{Bun}_G^{\mathrm{pt}}(X, x_0)$ of *pointed bundles* is the category whose objects are pairs (P, p_0) with p_0 a reference point in the fiber P_{x_0}, and whose morphisms are all morphisms of G-bundles.

In particular, for $X = \{*\}$ a point, the groupoid $G\text{-}\mathcal{Tor}^{\mathrm{pt}} := \mathcal{Bun}_G^{\mathrm{pt}}(*)$ is the groupoid of G-torsors with a reference point in the torsor.

LEMMA 1.67. The forgetful functor

$$\begin{aligned} \mathcal{Bun}_G^{\mathrm{pt}}(X, x_0) &\longrightarrow \mathcal{Bun}_G(X)\,, \\ (P, p_0) &\longmapsto P, \end{aligned} \tag{1.68}$$

which forgets the reference point p_0 is an equivalence of categories.

PROOF. The functor is fully faithful by definition, and it is essentially surjective because we are free to choose any point in the fiber P_{x_0} as reference point p_0. □

As a special case, $G\text{-}\mathcal{T}or^{\text{pt}}$ is equivalent to $G\text{-}\mathcal{T}or$. This allows us to define the functor $\widetilde{T}$ by the commutativity of the diagram

$$\begin{array}{ccc} \mathcal{B}un_G(X) & \xrightarrow{\ T\ } & [\Pi_1(X), G\text{-}\mathcal{T}or] \\ \cong\uparrow & & \downarrow \cong\ \iota_{x_0}^* \\ \mathcal{B}un_G^{\text{pt}}(X, x_0) & \xrightarrow[\widetilde{T}]{} & [*//\pi_1(X,x_0), G\text{-}\mathcal{T}or]. \end{array}$$

The functor $\widetilde{T}$ sends the G-bundle P with reference point p_0 to the functor

$$F: \quad *//\pi_1(X,x_0) \longrightarrow G\text{-}\mathcal{T}or$$

satisfying $F(*) = p_0$ and sends a closed path γ starting at x_0 to the endomorphism $T_\gamma^P : P_{x_0} \longrightarrow P_{x_0}$ of torsors which describes transport along the closed path γ. The reference point $p_0 \in P_{x_0}$ provides a lift $\widehat{T}$ of $\widetilde{T}$ via

$$\begin{array}{ccc} \mathcal{B}un_G^{\text{pt}}(X) & \xrightarrow{\ \widetilde{T}\ } & [*//\pi_1(X,x_0), G\text{-}\mathcal{T}or] \\ & \searrow^{\widehat{T}} & \uparrow \cong \\ & & [*//\pi_1(X,x_0), G\text{-}\mathcal{T}or^{\text{pt}}]. \end{array}$$

The following result gives an algebraic description of the groupoid in the bottom row of this diagram:

LEMMA 1.69. For H a finite group, the functor category $[*//H, G\text{-}\mathcal{T}or^{\text{pt}}]$ is canonically equivalent to the category $\text{Hom}(H,G)//G$ that has group homomorphisms $\varphi\colon H \longrightarrow G$ as objects and on which G acts by conjugation,

$$g.\varphi(h) := g\,\varphi(h)\,g^{-1}.$$

PROOF. The image $F(*)$ under the functor $F\colon *//H \longrightarrow G\text{-}\mathcal{T}or^{\text{pt}}$ is a G-torsor with reference point $p_0 \in F(*)$. For any morphism $h \in H$, $F(h)$ is a morphism of this G-torsor. The image $F(h)(p_0)$ of the reference point is of the form $\varphi(h).p_0$ for some $\varphi(h) \in G$. One checks that the so defined map $\varphi\colon H \longrightarrow G$ is a group homomorphism. This defines the map of objects for a functor

$$[*//H, G\text{-}\mathcal{T}or^{\text{pt}}] \longrightarrow \text{Hom}(H,G)//G\,.$$

A natural transformation $\alpha\colon F_1 \longrightarrow F_2$ in $[*//H, G\text{-}\mathcal{T}or^{\text{pt}}]$ has a single component $F_1(*) \longrightarrow F_2(*)$, which is a morphism of pointed G-torsors and which can thus be identified with an element $g_\alpha \in G$. The defining property of natural transformations then implies that $\varphi_2(h) = g_\alpha\,\varphi_1(h)\,g_\alpha^{-1}$. □

Composing the functor $\widehat{T}$ with the canonical equivalence from Lemma 1.69 (with $H = \pi_1(X,x_0)$) we get for any choice of base point $x_0 \in X$ a canonically defined functor

$$H: \quad \mathcal{B}un_G^{\text{pt}}(X,x_0) \longrightarrow \text{Hom}(\pi_1(X,x_0),G)//G \tag{1.70}$$

of groupoids. It sends a bundle $P \longrightarrow X$ with reference point $p_0 \in P_{x_0}$ to $\text{hol}_{p_0}^P$, the holonomy morphism (1.35):

$$\begin{aligned} \text{hol}: \quad \text{Obj}(\mathcal{B}un_G^{\text{pt}}(X,x_0)) &\longrightarrow \text{Hom}(\pi_1(X,x_0),G)\,, \\ (P,p_0) &\longmapsto \text{hol}_{p_0}^P\,. \end{aligned} \tag{1.71}$$

LEMMA 1.72. The functor (1.70) is an equivalence of categories.

PROOF. If H maps two morphisms Φ and $\Phi'\colon (P,p_0) \longrightarrow (P',p_0')$ to the same group element, then $\Phi'(p_0) = \Phi(p_0)$, so that $\Phi^{-1} \circ \Phi'$ is an automorphism preserving the reference point. Such an automorphism is necessarily the identity map. Moreover, for any $h \in G$ there exists a morphism Φ with $H(\Phi) = h$, namely the morphism $\Phi\colon P \longrightarrow P$ that in each fiber is acting by h. Thus H is fully faithful.

Furthermore, any group homomorphism $\varphi\colon \pi_1(X,x_0) \longrightarrow G$ defines a $\pi_1(X,x_0)$-action on G. Each orbit σ of this action gives a transitive group homomorphism from G to the symmetric group $S_{\ell(\sigma)}$ with $\ell(\sigma)$ the length of the orbit, and thereby an $\ell(\sigma)$-sheeted cover of X. The disjoint union, over all orbits, of these covers furnishes a pointed G-bundle P on X such that $\mathrm{hol}(P) = \varphi$. Thus the functor H is essentially surjective. The claim now follows from Lemma 1.62. □

This result implies that the other functors involved are equivalences of groupoids as well. For convenience we list all pertinent statements:

PROPOSITION 1.73.

(1) For any finite group G and any manifold X the functor

$$T: \quad \mathcal{Bun}_G(X) \longrightarrow [\Pi_1(X), G\text{-}\mathcal{Tor}]$$

that assigns to a G-bundle its transport operator is a canonical equivalence of categories.

(2) If X is a connected manifold, then any choice of base point $x \in X$ gives a canonical equivalence

$$\mathcal{Bun}_G^{\mathrm{pt}}(X) \xrightarrow{\simeq} \mathrm{Hom}(\pi_1(X,x), G)//G$$

between the category of pointed bundles and the action groupoid on the set of group homomorphisms.

(3) In particular, for any connected manifold X there is a bijection

$$\pi_0(\mathcal{Bun}_G(X)) \xrightarrow{\cong} \mathrm{Hom}(\pi_1(X,x), G)/G$$

between isomorphism classes of G-bundles on X and orbits of the adjoint G-action on $\mathrm{Hom}(\pi_1(X,x), G)$.

Later on we will make use of these results when investigating Dijkgraaf-Witten theories. We conclude this section by showing how they also allow us to make groupoids of bundles explicitly computable:

EXAMPLE 1.74. The fundamental group of the circle S^1 is freely generated by a path of winding number 1, and hence is isomorphic to the integers. It thus follows from Proposition 1.73(2) that the groupoid $\mathcal{Bun}_G(S^1)$ of G-bundles over the circle is equivalent to the groupoid $G//G$ of G acting on itself by conjugation. Explicitly, the set of objects of this groupoid is (noncanonically) the group G seen as a set, while a morphism between $g \in G$ and $g' \in G$ is a group element $h \in G$ such that $g' = hgh^{-1}$.

In other words, a G-bundle over the circle S^1 is determined by its holonomy around a generator of the fundamental group (with respect to any choice of base point) of the circle. The set $\pi_0(\mathcal{Bun}_G(S^1))$ of isomorphism classes of objects is the set of conjugacy classes in G, and the automorphism group $\mathrm{Aut}_{\mathcal{Bun}_G(S^1)}(P)$ of a G-bundle P with holonomy $g \in G$ is the centralizer group of the group element g.

As another application we derive an important finiteness property of the groupoid of bundles over a compact manifold.

DEFINITION 1.75. An *essentially finite category* is a category that is equivalent to a category whose class of morphisms (and hence in particular also its class of objects) is a finite set.

REMARK 1.76. Any essentially finite groupoid Γ is noncanonically equivalent to one of the form $\bigsqcup_{i=1}^{n} \mathcal{B}G_i$ with finite groups G_i. Indeed, choose representatives $x_1, x_2, \ldots, x_n \in \Gamma$ for the different isomorphism classes of Γ and set $G_i := \mathrm{Aut}_\Gamma(x_i)$. Then we obtain functors $\mathcal{B}G_i \longrightarrow \Gamma$ that send the single object of $\mathcal{B}G_i$ to x_i and are tautologically defined on morphisms. The induced functor $\bigsqcup_{i=1}^{n} \mathcal{B}G_i \longrightarrow \Gamma$ is an equivalence. (The reader is invited to check that the induced functor $\bigsqcup_{i=1}^{n} \mathcal{B}G_i \longrightarrow \Gamma$ is essentially surjective and fully faithful and hence is indeed an equivalence.)

COROLLARY 1.77. For any compact manifold X and any finite group G the groupoid $\mathcal{B}un_G(X)$ of G-bundles over X is essentially finite.

PROOF. Being compact, X has finitely many connected components; hence without loss of generality we can take X to be connected. In this case the assertion follows from Proposition 1.73(3) together with the facts that the fundamental group of a compact manifold is finitely generated [**Sak**, Lemma 1.2] and that there are only finitely many group homomorphisms from a finitely generated group to a finite group. (This exhibits $\mathcal{B}un_G(X)$ concretely as an essentially finite groupoid, albeit at the price of making a choice of the base point for the fundamental group, so that this realization is not canonical.) □

1.5. Limits and colimits

In this section we give a brief account of the theory of *limits* and *colimits* in categories, to the extent that is needed for our considerations. For a more comprehensive introduction we refer to [**Mac**] and to Chapter 11 of [**AdHS**].

DEFINITION 1.78. Let $\mathcal{I}$ be a small category and $\mathcal{C}$ an arbitrary category. A *diagram in $\mathcal{C}$ of shape $\mathcal{I}$* is a functor $X\colon \mathcal{I} \longrightarrow \mathcal{C}$.

EXAMPLE 1.79. Let us illustrate this concept by an easy concrete example. Let $\mathcal{I}$ be the category with three objects $\bullet$, $\star$, and $\diamond$ whose only nonidentity morphisms are a morphism $\alpha\colon \bullet \longrightarrow \star$ and a morphism $\beta\colon \diamond \longrightarrow \star$ (there are thus no compositions that need to be specified). We may depict $\mathcal{I}$ as follows:

$$\begin{array}{ccc} & & \diamond \\ & & \big\downarrow{\scriptstyle \beta} \\ \bullet & \xrightarrow[\alpha]{} & \star \end{array}$$

A diagram $X\colon \mathcal{I} \longrightarrow \mathcal{C}$ of shape $\mathcal{I}$ in a category $\mathcal{C}$ is completely specified by giving the three objects $Q := X(\bullet)$, $R := X(\star)$ and $S := X(\diamond)$ and the morphisms $f := X(\alpha)\colon Q \to R$ and $g := X(\beta)\colon S \to R$.

Diagrams of the shape described in Example 1.79, as well as dual versions of these, will be of particular interest to us. Accordingly we introduce separate terminology for them:

DEFINITION 1.80. A diagram in a category $\mathcal{C}$ of the form

$$\begin{array}{ccc} & & S \\ & & \downarrow{\scriptstyle g} \\ Q & \xrightarrow[f]{} & R \end{array} \tag{1.81}$$

is called a *cospan* in $\mathcal{C}$, while a diagram of the form

$$\begin{array}{ccc} & & S \\ & & \uparrow{\scriptstyle g} \\ Q & \xleftarrow[f]{} & R \end{array} \tag{1.82}$$

is called a *span* in $\mathcal{C}$.

EXAMPLE 1.83. The following class of diagrams is quite trivial, but nevertheless important. For any category $\mathcal{C}$, the (binary) *diagonal functor* $\Delta\colon \mathcal{C} \to \mathcal{C} \times \mathcal{C}$ is the functor that maps any object X to the pair (X, X) and any morphism f to the pair (f, f). More generally, for any small category $\mathcal{I}$, the *diagonal functor* is a functor Δ from $\mathcal{C}$ to the category $\mathcal{F}un(\mathcal{I}, \mathcal{C})$ of functors from $\mathcal{I}$ to $\mathcal{C}$. Δ assigns to the object $X \in \mathcal{C}$ the *constant functor* Δ_X, i.e., the diagram that maps every object in $\mathcal{I}$ to X and every morphism in $\mathcal{I}$ to id_X, and it assigns to a morphism f in $\mathcal{C}$ the natural transformation of diagrams whose component at every object in $\mathcal{I}$ is given by f. (The binary diagonal functor is recovered for $\mathcal{I}$ a category with two objects.)

DEFINITION 1.84. Let $\mathcal{I}$ be a small category and $\mathcal{C}$ be a category, and let $X\colon \mathcal{I} \to \mathcal{C}$ be a diagram in $\mathcal{C}$ of shape $\mathcal{I}$.

For every object $Y \in \mathcal{C}$, the constant functor gives such a diagram Δ_Y.

(1) A *cone to* the diagram $X\colon \mathcal{I} \to \mathcal{C}$ is an object $Y \in \mathcal{C}$ together with a natural transformation $\alpha\colon \Delta_Y \to X$ of functors $\mathcal{I} \to \mathcal{C}$.

    (Thus α is a family of morphisms $\alpha_i\colon Y \to X(i)$, one for each $i \in \mathcal{I}$, such that $\alpha_j = X(f) \circ \alpha_i$ for any morphism $f\colon i \to j$ in $\mathcal{I}$.)

(2) A *limit* of the diagram X is a cone $(Y, \alpha\colon \Delta_Y \to X)$ that is universal in the sense that for any cone $\beta\colon \Delta_Z \to X$ to X there is a unique morphism $\varphi\colon Z \to Y$ such that $\alpha \circ \Delta(\varphi) = \beta$.

    This property is referred to as the *universal property* of the limit.

REMARK 1.85. A limit is defined as a pair (Y, α), but it is common to suppress the *structure morphism* α in the notation and to refer to the object Y alone as a limit. Nonetheless, α is part of the data.

EXAMPLE 1.86. A cone to the cospan diagram from (1.81) is an object P in $\mathcal{C}$ together with morphisms $\pi_Q\colon P \longrightarrow Q$ and $\pi_S\colon P \longrightarrow S$ that make the diagram

(1.87)

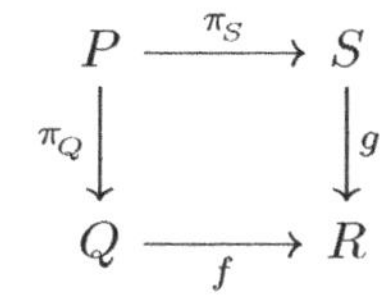

commute. This statement is extracted from Definition 1.84 as follows: The data of the cone consist of an object $P \in \mathcal{C}$ and three morphisms $\pi_Q\colon P \longrightarrow Q$, $\pi_S\colon P \longrightarrow S$ and $\pi_R\colon P \longrightarrow R$ subject to the conditions

$$f \circ \pi_Q = \pi_R \qquad \text{and} \qquad g \circ \pi_S = \pi_R\,.$$

Thus we can equivalently impose $f \circ \pi_Q = g \circ \pi_S$ and eliminate the redundant information π_R. The universality of this cone means that for *any* cone to the diagram (1.81), i.e., any object P' with morphisms $\pi'_Q\colon P' \longrightarrow Q$ and $\pi'_S\colon P' \longrightarrow S$ making the analog of (1.87) commute, there is a unique morphism $\varphi\colon P' \longrightarrow P$ such that $\pi_Q \circ \varphi = \pi'_Q$ and $\pi_S \circ \varphi = \pi'_S$. A convenient way to express this universality is by means of a diagram of the form

(1.88)

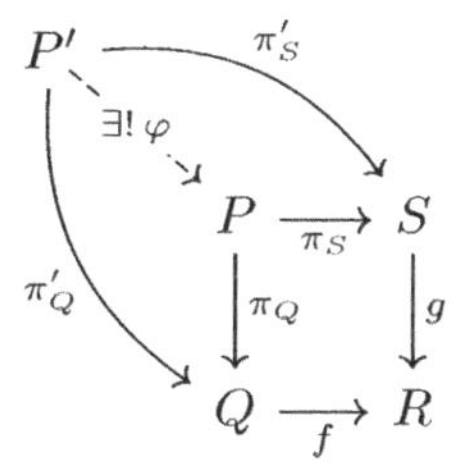

Such a type of limit is called a *pullback*.

We have in fact already encountered a special case of this structure: The pullback of a principal G-bundle is formed by taking a pullback in the category of manifolds; see Exercise 1.45.

The definition of a limit can be dualized, giving the notion of a colimit:

DEFINITION 1.89. Let $\mathcal{I}$ be a small category and $\mathcal{C}$ be a category, and let $X\colon \mathcal{I} \longrightarrow \mathcal{C}$ be a diagram in $\mathcal{C}$ of shape $\mathcal{I}$.

For every object $Y \in \mathcal{C}$, the constant functor gives such a diagram Δ_Y.

(1) A *cocone to* the diagram $X\colon \mathcal{I} \longrightarrow \mathcal{C}$ is an object $Y \in \mathcal{C}$ together with a natural transformation $\alpha\colon X \longrightarrow \Delta_Y$ of functors $\mathcal{I} \longrightarrow \mathcal{C}$.

   (Thus α is a family of morphisms $\alpha_i\colon X(i) \longrightarrow Y$, one for each $i \in \mathcal{C}$, such that $\alpha_j \circ X(f) = \alpha_i$ for any morphism $f\colon i \longrightarrow j$ in $\mathcal{I}$.)

(2) A *colimit* of the diagram X is a cocone $(Y, \alpha\colon X \longrightarrow \Delta_Y)$ that is universal in the sense that for any cocone $\beta\colon X \longrightarrow \Delta_Z$ from X there is a unique morphism $\varphi\colon Y \longrightarrow Z$ such that $\Delta(\varphi) \circ \alpha = \beta$.

According to Definition 1.84, a cone to an arbitrary diagram $X : \mathcal{I} \to \mathcal{C}$ consists of a object Y in $\mathcal{C}$ and morphisms $\alpha_i : Y \to X(i)$ for all $i \in \mathcal{I}$ such that for every morphism $f : i \to j$ in $\mathcal{I}$ the triangle

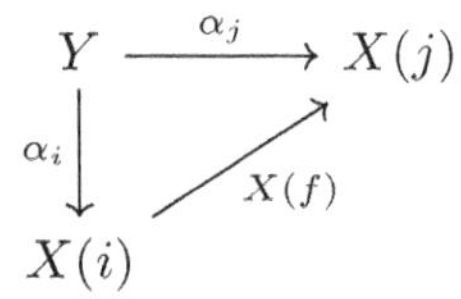

commutes. Generalizing (1.88), the diagrammatic representation of the universality of the cone (Y, α) is expressed as follows:

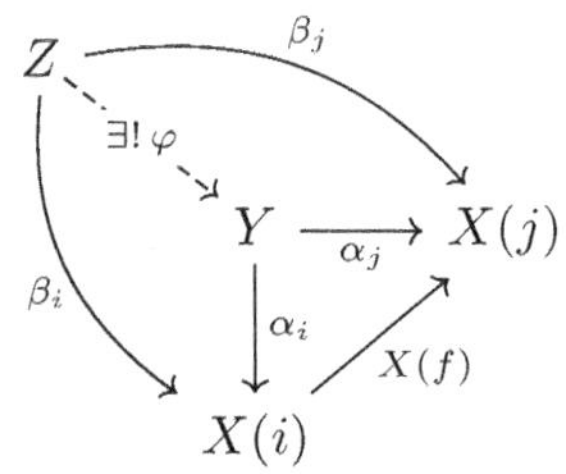

LEMMA 1.90. A limit of a diagram, if any exists, is unique up to canonical isomorphism. More specifically, assume that both $\alpha : Y \to X$ and $\alpha' : Y' \to X$ are limits to the same diagram X. Then there is a unique isomorphism $\varphi : Y \to Y'$ such that $\alpha' \circ \varphi = \alpha$.

PROOF. Since $\alpha : Y \to X$ is a cone to X, by the universal property for the limit $\alpha' : Y' \to X$ there is a unique morphism $\varphi : Y \to Y'$ such that $\alpha' \circ \varphi = \alpha$. It remains to show that the so defined morphism φ is invertible. To this end, note that by exchanging the roles of α and α' we find a unique morphism $\widetilde{\varphi} : Y' \to Y$ such that $\alpha \circ \widetilde{\varphi} = \alpha'$. This entails that the morphism $\widetilde{\varphi} \circ \varphi : Y \to Y$ satisfies the equalities $\alpha \circ \widetilde{\varphi} \circ \varphi = \alpha' \circ \varphi = \alpha$. By the universal property of α this implies, in turn, that $\widetilde{\varphi} \circ \varphi = \mathrm{id}_Y$. After exchanging once again the roles of α and α' we find likewise that $\varphi \circ \widetilde{\varphi} = \mathrm{id}_{Y'}$. □

This result justifies to refer to *a* (co)limit of a diagram X, if it exists, as *the* (co)limit. The (co)limit of X is often denoted by $\lim X$ and $\operatorname{colim} X$, respectively.

EXAMPLE 1.91 (Products and coproducts). Recall from Example 1.5(1) that every set I may be regarded as a discrete category, having only identity morphisms. A diagram X of shape I in a category $\mathcal{C}$ is nothing but a family of objects indexed over I, i.e., $X = (X_i)_{i \in I}$. The (co)limit over the I-shaped diagram $X = (X_i)_{i \in I}$ in $\mathcal{C}$ (if it exists) is called the *(co)product over the family* $(X_i)_{i \in I}$. For $\mathcal{C} = \mathcal{S}et$ the category of sets, one may deduce from the universal property of the limit that the product of $(X_i)_{i \in I}$ (in the categorical sense just defined) is the *Cartesian product* $\prod_{i \in I} X_i$ together with the projections $p_i : \prod_{j \in I} X_j \to X_i$. Indeed, these projection maps endow the Cartesian product $\prod_{i \in I} X_i$ with the structure of a cone to the diagram $(X_i)_{i \in I}$ (there is nothing to check because the shape category is discrete). The universality of this cone amounts to the statement that any family $\beta_i : Z \to X_i$

of maps of sets induces a map (which by Lemma 1.90 is unique) $B\colon Z \longrightarrow \prod_{i\in I} X_i$ such that $p_i \circ B = \beta_i$, namely the map sending $z \in Z$ to $(\beta_i(z))_{i\in I}$. In other words, the assignment $(\beta_i)_{i\in I} \mapsto B$ just described yields a bijection

$$\prod_{i\in I} \mathrm{Hom}_{\mathcal{S}et}(Z, X_i) \cong \mathrm{Hom}_{\mathcal{S}et}\Big(Z, \prod_{i\in I} X_i\Big). \tag{1.92}$$

The coproduct over $(X_i)_{i\in I}$ is the *disjoint union* $\coprod_{i\in I} X_i$ together with the inclusions $\iota_i\colon X_i \longrightarrow \coprod_{j\in I} X_j$. The universal property can be verified using an argument dual to the one just given for the product. It relies on the canonical bijection

$$\prod_{i\in I} \mathrm{Hom}_{\mathcal{S}et}\Big(X_i, Z\Big) \cong \mathrm{Hom}_{\mathcal{S}et}\Big(\coprod_{i\in I} X_i, Z\Big) \tag{1.93}$$

that is induced by the inclusion maps $\iota_i\colon X_i \longrightarrow \coprod_{i\in I} X_i$. (The isomorphisms (1.92) and (1.93) will be reinterpreted in Exercise 1.132.)

EXERCISE 1.94. Show that in the category of sets and in the category of topological spaces both products and coproducts exist, and that the product is given by the Cartesian product while the coproduct is given by disjoint union.

Further show that products and coproducts exist in the category of vector spaces over a field, and that they are given by products and direct sums of vector spaces, respectively.

Under which assumption on the indexing set are products and coproducts in these categories isomorphic?

DEFINITION 1.95. A category in which limits (colimits) for all diagrams over small categories exist is called *complete* (*cocomplete*).

Instead of asking for *all* (co)limits to exist, it often makes sense to just ask for the existence of a certain type of (co)limit. For example, we say that a category *has (co)products* if all (co)products exist in the category. Next we study specific types of (co)limits.

Consider a diagram, in a category $\mathcal{C}$, that has the shape of a category with two objects and two parallel nonidentity morphisms. Such a diagram is determined by the choice of two objects in $\mathcal{C}$ and two parallel morphisms in $\mathcal{C}$. It may thus be depicted as

$$X \underset{g}{\overset{f}{\rightrightarrows}} Y. \tag{1.96}$$

DEFINITION 1.97. A (co)limit (if it exists) of a diagram of the form (1.96) is called the *(co)equalizer* of the parallel pair (f, g). It is denoted by $\mathrm{eq}(f, g)$ and $\mathrm{coeq}(f, g)$, respectively.

EXAMPLE 1.98. For $\mathcal{C} = \mathcal{S}et$ the category of sets, the equalizer of (f, g) can be described as

$$\mathrm{eq}(f, g) = \{x \in X \mid f(x) = g(x)\},$$

where the structure maps are the inclusion $\mathrm{eq}(f, g) \subseteq X$ and the restriction

$$f|_{\mathrm{eq}(f,g)}\colon \quad \mathrm{eq}(f, g) \subseteq X \xrightarrow{f} Y$$

of f to $\mathrm{eq}(f, g)$, which by definition agrees with the restriction of g to $\mathrm{eq}(f, g)$.

The universal property amounts to the statement that for any map $h\colon Z \to X$ with $f\circ h = g\circ h$ there is a map (which, again by Lemma 1.90, is unique)

$$\varphi:\quad Z \longrightarrow \mathrm{eq}(f,g) = \{x\in X \mid f(x) = g(x)\} \longrightarrow Y$$

such that the equality $f|_{\mathrm{eq}(f,g)}\circ\varphi = g|_{\mathrm{eq}(f,g)}\circ\varphi$ holds, namely the map given by $\varphi(z) = h(z)$ for $z\in Z$. (The distinction between φ and h is merely that φ is regarded as a map with values in $\mathrm{eq}(f,g)$.)

The coequalizer of (f,g) can be described as

$$\mathrm{coeq}(f,g) = Y/\!\sim,$$

where $\sim$ is the equivalence relation on Y which, for any $x\in X$, declares the two elements $f(x)$ and $g(x)$ of Y to be equivalent. The structure morphisms are the maps $X \xrightarrow{f} Y \to \mathrm{coeq}(f,g)$, that is, the composition of $f\colon X\to Y$ with the quotient map $Y\to\mathrm{coeq}(f,g)$ (again, using g instead of f yields the same result). The universal property is verified by an analogous argument that is dual to the one for equalizers, in the sense (compare Definition 1.13(1)) that the direction of all arrows is reversed.

EXAMPLE 1.99. The tensor product $M\otimes_R N$ of two bimodules M and N over a ring R is the coequalizer of the diagram

$$M\times R\times N \underset{\mathrm{id}_M\times\rho}{\overset{\mathfrak{q}\times\mathrm{id}_N}{\rightrightarrows}} M\times N\,,$$

where $\mathfrak{q}$ is the right action of R on M and ρ the left action of R on N.

EXERCISE 1.100. Prove that the category of topological spaces and the category of modules over a ring have equalizers and coequalizers. Describe them.

PROPOSITION 1.101. *If a category admits all products and all equalizers, then it is complete in the sense of Definition 1.95.*

Dually, if a category admits all coproducts and coequalizers, then it is cocomplete.

SKETCH OF PROOF. We provide a formula for an arbitrary limit in terms of products and equalizers and leave it to the reader to verify the universal property.

The limit of a diagram $X\colon \mathcal{I}\to\mathcal{C}$, if it exists, can be expressed as the equalizer

$$\lim X = \mathrm{eq}\Big(\prod_{i\in\mathcal{I}} X(i) \underset{\mu}{\overset{\lambda}{\rightrightarrows}} \prod_{f\colon i\to j \text{ in } \mathcal{I}} X(j)\Big)\,.$$

Here the maps λ and μ are defined with the help of the universal property of the product as follows: The components $\lambda^f,\ \mu^f\colon \prod_{i\in I} X(i)\to X(j)$ corresponding to a morphism $f\colon i\to j$ in $\mathcal{I}$ are given by

$$\lambda^f:\quad \prod_{k\in I} X(k) \xrightarrow{\text{projection}_i} X(i) \xrightarrow{X(f)} X(j)$$

and

$$\mu^f:\quad \prod_{k\in I} X(k) \xrightarrow{\text{projection}_j} X(j)\,,$$

respectively. □

EXERCISE 1.102. Use Proposition 1.101 to conclude that each of the categories of sets, of topological spaces, and of modules over a ring, is both complete and cocomplete.

Show that the category of manifolds is neither complete nor cocomplete.

(*Hint*: Show that pullbacks do not necessarily exist.)

EXERCISE 1.103. A representation of a group G on a vector space V may be regarded as a functor $*/\!/G \longrightarrow \mathcal{V}ect(\mathbb{K})$ which sends $*$ to V.

Compute the limit and colimit of $*/\!/G \longrightarrow \mathcal{V}ect(\mathbb{K})$ and prove that both are isomorphic if G is finite and $\mathbb{K}$ has characteristic zero.

REMARK 1.104. The vector spaces computed in Exercise 1.103 can be described as follows. The *invariants* V^G of a G-representation V are given by the vector space of fixed points of the G-action, i.e.

$$V^G := \{v \in V \mid g.v = v \text{ for every } g \in G\},$$

or, in other words, by the largest trivial subrepresentation of V. Dually, the *coinvariants* V_G of V are defined as the quotient vector space

$$V_G = V/(g.v - v \mid g \in G, v \in V),$$

i.e., as the largest quotient with trivial G-action.

As explained in Example 1.86, a limit of a cospan

$$\begin{array}{ccc} & & Z \\ & & \downarrow{\scriptstyle g} \\ X & \xrightarrow[f]{} & Y \end{array} \tag{1.105}$$

is called a pullback. Concerning notation, it is common to write the pullback (1.105) also as $X \times_Y Z$; the commuting square (1.87) that exhibits it as a limit then takes the form

$$\begin{array}{ccc} X \times_Y Z & \xrightarrow{\pi_X} & Z \\ {\scriptstyle \pi_Z}\downarrow & & \downarrow{\scriptstyle g} \\ X & \xrightarrow[f]{} & Y. \end{array}$$

Similarly, a colimit of a span

$$\begin{array}{ccc} Y & \xrightarrow{f} & X \\ {\scriptstyle g}\downarrow & & \\ Z & & \end{array} \tag{1.106}$$

is called a *pushout*, and is often denoted by $X \cup_Y Z$. Together with the structure maps that exhibit it as a colimit, a pushout forms a commuting square

$$\begin{array}{ccc} Y & \xrightarrow{f} & X \\ {\scriptstyle g}\downarrow & & \downarrow \\ Z & \longrightarrow & X \cup_Y Z. \end{array} \tag{1.107}$$

EXERCISE 1.108.

(1) Prove that in the category of sets a pullback of a diagram of the shape (1.105) is given by the set
$$X\times_Y Z=\{(x,z)\in X\times Z\mid f(x)=g(z)\}\,.$$
(2) Show that the pullback of bundles from Exercise 1.45 is a pullback in the category of manifolds.

(3) Prove that a pushout in the category of sets of a diagram of the shape (1.106) is given by the set
$$X\cup_Y Z=(X\amalg Z)/\!\sim\,,$$
where the relation $\sim$ is defined to be the smallest equivalence relation that satisfies $f(y)\sim g(y)$ for all $y\in Y$.

(4) Describe pullbacks and pushouts in the category of modules over a ring and in the category of topological spaces.

(5) How can pullbacks and pushouts be written as a combination of (co)products and (co)equalizers?

Hint: Use Proposition 1.101 and the sketch for its proof. If you still get stuck, consult a textbook, e.g., [**Ri2**] or [**AdHS**].

DEFINITION 1.109. A *terminal object* in a category $\mathcal{C}$ is an object $Z\in\mathcal{C}$ such that for any object $Y\in\mathcal{C}$ there is precisely one morphism with domain Y and codomain Z. Dually, an *initial object* in $\mathcal{C}$ is an object $X\in\mathcal{C}$ such that for any $Y\in\mathcal{C}$ there is precisely one morphism from X to Y. An object that is both initial and terminal is called a *zero object*, and is commonly denoted by $0\in\mathcal{C}$.

If a terminal or initial object exists, then it is unique up to unique isomorphism, and it can be identified with the empty product and the empty coproduct in the category, respectively. If a zero object 0 exists, then for any pair $X,Y\in\mathcal{C}$ of objects there is a distinguished morphism $X\longrightarrow Y$, namely the composite $X\longrightarrow 0\longrightarrow Y$; this is called the *zero morphism* in $\mathrm{Hom}(X,Y)$.

EXAMPLE 1.110. The category of sets has an initial object given by the empty set, and a terminal object given by the *singleton*, i.e., the set with one element; in particular it does not have a zero object. In the category of vector spaces, the zero vector space $\{0\}$ is both an initial and a terminal object, and thus a zero object. The zero morphism between two vector spaces V and W is the linear map that sends every $v\in V$ to $0\in W$.

DEFINITION 1.111. Let $\mathcal{C}$ be a category with zero object. The *biproduct*, also called the *direct sum*, of a finite (possibly empty) collection of objects $X_1,X_2,\dots,$ X_n in $\mathcal{C}$ is an object $X_1\oplus X_2\oplus\cdots\oplus X_n$ in $\mathcal{C}$ endowed with structure morphisms $p_k\colon X_1\oplus X_2\oplus\cdots\oplus X_n\longrightarrow X_k$ (which are epimorphisms and are called projection morphisms) and $\imath_k\colon X_k\longrightarrow X_1\oplus X_2\oplus\cdots\oplus X_n$ (which are monomorphisms and are called embedding morphisms) for $k=1,2,\dots,n$, such that the collection $\{p_k\}$ of projection morphisms turns the object $X_1\oplus X_2\oplus\cdots\oplus X_n$ into a product and the collection $\{\imath_k\}$ of embedding morphisms turns it into a coproduct, and such that the compatibility relations
$$p_k\circ\imath_k=\mathrm{id}_{X_k}\qquad\text{and}\qquad p_k\circ\imath_l=0\ \in\mathrm{Hom}(X_l,X_k)\qquad\text{for } k\neq l$$
are fulfilled.

DEFINITION 1.112. For a category $\mathcal{C}$ with zero object, the *kernel* of a morphism $f\colon X \to Y$ in $\mathcal{C}$ is the pullback

$$\begin{array}{ccc} \ker f & \longrightarrow & X \\ \downarrow & & \downarrow f \\ 0 & \longrightarrow & Y \end{array}$$

provided that this pullback exists.

Similarly, the *cokernel* of f is the pushout

$$\begin{array}{ccc} X & \xrightarrow{f} & Y \\ \downarrow & & \downarrow \\ 0 & \longrightarrow & \operatorname{coker} f, \end{array}$$

again provided that this pushout exists.

DEFINITION 1.113. An *additive* category is a category with the following properties:

- Every morphism set has the structure of an abelian group or, in other words, of a $\mathbb{Z}$-module (recall Example 1.5(4)), in such a way that the composition is $\mathbb{Z}$-bilinear.
- All finite products and finite coproducts exist.

EXERCISE 1.114. Show that every additive category has a zero object and that finite products and finite coproducts in an additive category coincide up to canonical isomorphism.

EXAMPLE 1.115. Examples for categories that are not additive are the category of sets (see Example 1.110), the category of unital rings, which has $\mathbb{Z}$ as initial object and the one-element set as terminal object, and the category of groups, in which the product is the Cartesian product, while the coproduct is the free product of groups.

DEFINITION 1.116. An *abelian* category is an additive category with the following additional properties:

- Every morphism has both a kernel and a cokernel.
- For any monomorphism $f\colon X \to Y$ the square

$$\begin{array}{ccc} X & \xrightarrow{f} & Y \\ \downarrow & & \downarrow \\ 0 & \longrightarrow & \operatorname{coker} f \end{array}$$

  (which by definition is a pushout) is a pullback; in short, "every monomorphism is the kernel of its cokernel".

- For any epimorphism $f\colon X \to Y$ the square

$$\begin{array}{ccc} \ker f & \longrightarrow & X \\ \big\downarrow & & \big\downarrow{\scriptstyle f} \\ 0 & \longrightarrow & Y \end{array}$$

(which by definition is a pullback) is a pushout; in short, "every epimorphism is the cokernel of its kernel".

EXERCISE 1.117. Show that the category of modules over any ring R, and hence in particular the category of vector spaces over a field $\mathbb{K}$, is abelian.

Hint: In this case the kernel of a map $f\colon X \longrightarrow Y$ of modules over R is given by the customary point-set expression $\ker f = \{x \in X \mid f(x) = 0\}$ for the kernel.

REMARK 1.118. The latter example is in fact prototypical: According to the *Freyd-Mitchell embedding theorem*, every abelian category is a full subcategory of a category of modules over some ring, in such a way that the embedding functor is an exact functor. For details, see, e.g., Chapter 4.4 of [**Frey**].

EXAMPLE 1.119. Examples for categories that are additive but not abelian are the category of Hilbert spaces, the category of even-dimensional vector spaces, and the category of free abelian groups.

EXERCISE 1.120. For a ring R, an R-module P is called *projective* if for any epimorphism $q\colon X \longrightarrow Y$ of R-modules and any morphism $f\colon P \longrightarrow Y$ of R-modules there is a morphism $\widetilde{f}\colon P \longrightarrow X$ of R-modules such that the diagram

$$\begin{array}{ccc} & & X \\ & \overset{\widetilde{f}}{\nearrow} & \big\downarrow{\scriptstyle q} \\ P & \xrightarrow[f]{} & Y \end{array}$$

commutes.

Show that the full subcategory of the abelian category of R-modules that has as objects the projective R-modules is additive, but in general not abelian.

Hint: The projective modules in the category of finitely generated $\mathbb{Z}$-modules are the free abelian groups.

DEFINITION 1.121. An object X in an abelian category is called *simple* if 0 and X are its only subobjects and $X \neq 0$.

An abelian category $\mathcal{C}$ is called *semisimple* if every object is isomorphic to a coproduct of simple objects.

REMARK 1.122. Let M be a simple module over a ring R, and X an arbitrary R-module. Then *Schur's lemma* states that a module morphism $f\colon M \longrightarrow X$ is either injective or zero, and a module morphism $g\colon X \longrightarrow M$ is either surjective or zero. It follows in particular that any module morphism between two simple R-modules is either an isomorphism or zero. These statements follow from the fact that the kernel $\ker(f) \subseteq M$ of f and the image $g(X) \subseteq M$ of g are submodules of M and thus, M being simple, can only be either all of M or 0.

Schur's lemma can also be formulated in categorical terms: Let $\mathcal{C}$ be an abelian category, $S \in \mathcal{C}$ a simple object, and X an arbitrary object in $\mathcal{C}$. Then a morphism

$S \to X$ is either a monomorphism or zero, and a morphism $X \to S$ is either an epimorphism or zero, and hence in particular any morphism between simple objects is either an isomorphism or zero.

There is also a generalization to categories that are not abelian, like $\mathbb{K}$-linear categories of induced modules over graded algebras. It states that the vector space of morphisms between any two simple objects X and Y is either zero or isomorphic, as a graded vector space, to the spaces of endomorphisms of both X and Y, but the latter spaces contain, in general, noninvertible morphisms.

REMARK 1.123. In the $\mathbb{K}$-linear setting, besides simplicity also the following notion is of interest: An object X in a $\mathbb{K}$-linear category is called *absolutely simple* if its endomorphism algebra is isomorphic to the ground field, i.e., if

$$\mathrm{Hom}(X,X) = \mathbb{K}\,\mathrm{id}_X\,. \tag{1.124}$$

The two notions "simple" and "absolutely simple" are distinct, but closely related: If $\mathbb{K}$ is algebraically closed (as we generally assume) and $\mathcal{C}$ is $\mathbb{K}$-linear abelian, then $X \in \mathcal{C}$ being simple implies that X is absolutely simple. Conversely, if $\mathcal{C}$ is $\mathbb{K}$-linear, with $\mathbb{K}$ any field, and has a zero object, and every monomorphism $m\colon X \to Y$ in $\mathcal{C}$ is split (meaning that there exists a morphism $r\colon Y \to X$ such that $r \circ m = \mathrm{id}_X$), then $X \in \mathcal{C}$ being absolutely simple implies that X and 0 are the only subobjects of X, so that X is, if nonzero, simple in the sense of Definition 1.121 (without necessarily assuming that $\mathcal{C}$ is abelian).

The following finite variant of the notion of semisimple category will also be important to us:

DEFINITION 1.125. An abelian category $\mathcal{C}$ is called *finitely semisimple* if $\mathcal{C}$ has only finitely many simple objects up to isomorphism and if every object is isomorphic to a finite coproduct (and hence finite product) of simple objects.

EXERCISE 1.126. Prove that the category of finite-dimensional vector spaces over a field $\mathbb{K}$ is finitely semisimple.

1.6. Adjoint functors

There is one further key concept of category theory that we will need: a functor can admit an *adjoint functor*, which is a functor with interchanged domain and codomain.

DEFINITION 1.127. Let $\mathcal{C}$ and $\mathcal{D}$ be categories and F and G be functors $F\colon \mathcal{C} \to \mathcal{D}$ and $G\colon \mathcal{D} \to \mathcal{C}$. Then F is called a *left adjoint* functor of G, and G a *right adjoint* functor of F, in symbols $F \dashv G$, if there are natural transformations

$$\varepsilon: \quad F \circ G \longrightarrow \mathrm{id}_{\mathcal{D}} \qquad \text{and} \qquad \eta: \quad \mathrm{id}_{\mathcal{C}} \longrightarrow G \circ F$$

such that for all objects c in $\mathcal{C}$ and d in $\mathcal{D}$ the morphisms

$$G(d) \xrightarrow{\eta_{G(d)}} \big(G{\circ}F\big) \circ G(d) = G \circ (F{\circ}G)(d) \xrightarrow{G(\varepsilon_d)} G(d)$$

and

$$F(c) \xrightarrow{F(\eta_c)} F \circ (G{\circ}F)(c) = (F{\circ}G) \circ F(c) \xrightarrow{\varepsilon_{F(c)}} F(c)$$

are identities.

One also says that one deals with an *adjunction* $F \dashv G$ and refers to ε and η as the *counit* and *unit* of the adjunction, respectively.

If F is both a left and a right adjoint of G, then it is called a *two-sided adjoint* and the adjunction $F \dashv G \dashv F$ is called *ambidextrous*.

An equivalent definition can be given as follows:

DEFINITION 1.128. Let $\mathcal{C}$ and $\mathcal{D}$ be categories. A functor $F\colon \mathcal{C} \to \mathcal{D}$ is called *left adjoint* to a functor $G\colon \mathcal{D} \to \mathcal{C}$ if for any pair of objects c in $\mathcal{C}$ and d in $\mathcal{D}$ there is an isomorphism

$$\Phi_{c,d}: \quad \mathrm{Hom}_{\mathcal{C}}(c, G(d)) \xrightarrow{\cong} \mathrm{Hom}_{\mathcal{D}}(F(c), d)$$

of morphism sets with the following naturality property:

Given a pair of morphisms $f\colon c' \to c$ in $\mathcal{C}$ and $g\colon d \to d'$ in $\mathcal{D}$, form for every morphism $\varphi \in \mathrm{Hom}_{\mathcal{D}}(F(c), d)$ the composite

$$\mathrm{Hom}(F(f), g)(\varphi) := \left(F(c') \xrightarrow{F(f)} F(c) \xrightarrow{\varphi} d \xrightarrow{g} d'\right)$$

in $\mathrm{Hom}_{\mathcal{D}}(F(c'), d')$, and for every $\psi \in \mathrm{Hom}_{\mathcal{C}}(c, G(d))$ the morphism

$$\mathrm{Hom}(f, G(g))(\psi) := \left(c' \xrightarrow{f} c \xrightarrow{\psi} G(d) \xrightarrow{G(g)} G(d')\right)$$

in $\mathrm{Hom}_{\mathcal{C}}(c', G(d'))$. Then it is required that the diagram

$$\begin{array}{ccc}
\mathrm{Hom}_{\mathcal{C}}(c, G(d)) & \xrightarrow{\mathrm{Hom}(f,G(g))} & \mathrm{Hom}_{\mathcal{C}}(c', G(d')) \\
{\scriptstyle \Phi_{c,d}}\downarrow & & \downarrow{\scriptstyle \Phi_{c',d'}} \\
\mathrm{Hom}_{\mathcal{D}}(F(c), d) & \xrightarrow{\mathrm{Hom}(F(f),g)} & \mathrm{Hom}_{\mathcal{D}}(F(c'), d')
\end{array}$$

commutes for all morphisms f, g.

Thinking of $\mathrm{Hom}(-,-)$ as a scalar product, this version of the definition explains the term "adjoint".

Let us sketch the relation between Definitions 1.127 and 1.128 (we encourage the reader to work out some details and refer for a proof to Chapter IV of [**Mac**]):

(1) Let $F \dashv G$ be adjoint functors, defined according to Definition 1.128. Then we are in particular given isomorphisms

$$\mathrm{Hom}_{\mathcal{C}}(G(d), G(d)) \xrightarrow{\cong} \mathrm{Hom}_{\mathcal{D}}(F(G(d)), d)$$

and

$$\mathrm{Hom}_{\mathcal{D}}(F(c), F(c)) \xrightarrow{\cong} \mathrm{Hom}_{\mathcal{C}}(c, G(F(c)))$$

for all $c \in \mathcal{C}$ and all $d \in \mathcal{D}$. The images of the identity morphisms on $G(d)$ and on $F(c)$, respectively, under these isomorphisms are the components of natural transformations

$$\varepsilon: \quad F \circ G \longrightarrow \mathrm{id}_{\mathcal{D}} \qquad \text{and} \qquad \eta: \quad \mathrm{id}_{\mathcal{C}} \longrightarrow G \circ F\,.$$

These satisfy the relations for the unit and counit required in Definition 1.127.

(2) Conversely, we can recover the adjunction isomorphisms $\Phi_{c,d}$ from the natural transformations ε and η by composing

$$\mathrm{Hom}_{\mathcal{C}}(c, G(d)) \xrightarrow{F} \mathrm{Hom}_{\mathcal{D}}(F(c), F(G(d)) \xrightarrow{(\varepsilon_d)_*} \mathrm{Hom}_{\mathcal{D}}(F(c), d)\,,$$

where $(\varepsilon_d)_*$ is the map that is obtained by postcomposition with the morphism $\varepsilon_d\colon F\circ G(d) \longrightarrow d$ on the morphism set, and their inverses by

$$\mathrm{Hom}_{\mathcal{D}}(F(c), d) \xrightarrow{G} \mathrm{Hom}_{\mathcal{C}}(G(F(c)), G(d)) \xrightarrow{\eta_d^*} \mathrm{Hom}_{\mathcal{C}}(c, G(d))\,,$$

with η_d^* acting by precomposition.

(3) A pair $F \dashv G$ of adjoint functors is an equivalence of categories if and only if ε and η are natural isomorphisms of functors.

EXERCISE 1.129. Use the Yoneda Lemma from Example 1.55 to show that for any functor a left and right adjoint, respectively, is unique up to natural isomorphism if it exists. (It even suffices to just determine either the unit or the counit of the adjunction; see, e.g., Remark 4.2.7 in [**Ri2**].)

This allows us to speak of *the* left (right) adjoint of a functor, provided that it exists.

EXERCISE 1.130. Derive Lemma 1.90 from the statement in Exercise 1.129.

EXAMPLES 1.131.

(1) The forgetful functor

$$U: \quad \mathcal{V}ect(\mathbb{K}) \longrightarrow \mathcal{S}et$$

which assigns to any $\mathbb{K}$-vector space the underlying set has as a left adjoint the functor

$$F: \quad \mathcal{S}et \longrightarrow \mathcal{V}ect(\mathbb{K})$$

that associates to a set X the freely generated vector space on X, i.e., $F(X)$ is the $\mathbb{K}$-vector space with basis X. To see that F is left adjoint to U, note that for any set M and any $\mathbb{K}$-vector space V we have an isomorphism

$$\begin{aligned} \Phi_{M,V}: \quad \mathrm{Hom}_{\mathcal{S}et}(M, U(V)) &\longrightarrow \mathrm{Hom}_{\mathbb{K}}(F(M), V)\,, \\ \varphi &\longmapsto \Phi_{M,V}(\varphi)\,, \end{aligned}$$

where $\Phi_{M,V}(\varphi)$ is the $\mathbb{K}$-linear map that is obtained by prescribing values in V on the basis M of $F(M)$ using φ and extending this prescription linearly:

$$\Phi_{M,V}(\varphi)\Big(\sum_{m\in M} \lambda_m\, m\Big) := \sum_{m\in M} \lambda_m\, \varphi(m)\,.$$

In particular, we find an isomorphism

$$\mathrm{Hom}_{\mathcal{S}et}(\emptyset, U(V)) \cong \mathrm{Hom}_{\mathbb{K}}(F(\emptyset), V)$$

of sets for every $\mathbb{K}$-vector space V. Thus $\mathrm{Hom}_{\mathbb{K}}(F(\emptyset), V)$ has exactly one element for any vector space V. This shows that $F(\emptyset) = \{0\}$, i.e., the vector space freely generated by the empty set is the zero-dimensional vector space, which is the initial object (and hence, $\mathcal{V}ect(\mathbb{K})$ being abelian, the zero object) in the category of vector spaces.

(2) More generally, freely generated objects are obtained as images under left adjoints of forgetful functors. It is, however, not true that every forgetful functor admits a left adjoint. As a counterexample, take the forgetful functor U from the category $\mathcal{F}ield$ of fields to the category $\mathcal{S}et$ of sets. Suppose a left adjoint of U exists; then the image K of the empty set under the left adjoint must be a field such that for any other field L there is a bijection

$$\mathrm{Hom}_{\mathcal{F}ield}(K, L) \cong \mathrm{Hom}_{\mathcal{S}et}(\emptyset, U(L)) \cong \star\,,$$

with $\star$ denoting the one-element set. Now any nonzero morphism of fields is injective. As a consequence, a field K with this property would be a subfield of any field L. Such a field does not exist.

EXERCISE 1.132. Recall the diagonal functor Δ from Example 1.83. Show that Δ has the coproduct (or direct sum) as a left adjoint and the product as a right adjoint.

EXAMPLE 1.133. Let R and S be rings and let $\phi\colon R \longrightarrow S$ be a ring homomorphism. As seen in Example 1.42(6), the pullback, or restriction of scalars, provides a functor

$$\phi^* : \quad S\text{-mod} \longrightarrow R\text{-mod}\,.$$

Such a functor arises, e.g., from a homomorphism $\iota\colon H \longrightarrow G$ of groups when considering the group algebras $R := \mathbb{K}[H]$ and $S := \mathbb{K}[G]$.

Let us compute its adjoint functors.

(1) A left adjoint $\phi_!\colon R\text{-mod} \longrightarrow S\text{-mod}$ is obtained by defining it on objects as the tensor product over R,

$$\phi_!(M) := S \otimes_R M\,,$$

where S is a right R-module by pullback along ϕ and where the left S-action is given by left multiplication

$$s'.\,(s \otimes m) := (s's) \otimes m\,.$$

On morphisms, we set

$$\phi_!(f) := \mathrm{id}_S \otimes_R f\,.$$

The so defined functor $\phi_!$ is called *extension of scalars* or *induction.*

To see that $\phi_!$ is indeed left adjoint to ϕ^*, consider for $M \in R$-mod and $N \in S$-mod the following two morphisms of abelian groups:

$$\begin{aligned} \mathrm{Hom}_R(M, \phi^*(N)) &\longrightarrow \mathrm{Hom}_S(S \otimes_R M, N)\,,\\ f &\longmapsto \big(s \otimes m \mapsto s.f(m)\big) \end{aligned}$$

and

$$\begin{aligned} \mathrm{Hom}_S(S \otimes_R M, N) &\longrightarrow \mathrm{Hom}_R(M, \phi^*(N))\,,\\ g &\longmapsto \big(m \mapsto g(1_S \otimes m)\big)\,. \end{aligned}$$

It is not hard to verify that these morphisms are mutually inverse, and that they possess the naturality property of adjoint functors from Definition 1.128.

(2) To obtain a right adjoint $\phi_*\colon R\text{-mod} \longrightarrow S\text{-mod}$ for the pullback functor ϕ^*, we define a functor ϕ_* on objects by

$$\phi_*(M) := \mathrm{Hom}_R(S, M)\,.$$

Here the ring S is regarded as a left R-module by pullback along the ring homomorphism ϕ, whereby $\phi^*(M)$ becomes a left S-module by the right action of S on itself: $(s.\varphi)(s') := \varphi(s's)$ for $\varphi \in \mathrm{Hom}_R(S, M)$. The functor ϕ^* is called *coinduction*.

Again we consider for $M \in R$-mod and $N \in S$-mod two morphisms of abelian groups:

$$\begin{aligned} \mathrm{Hom}_R(\phi^*(N), M) &\longrightarrow \mathrm{Hom}_S(N, \mathrm{Hom}_R(S, M))\,, \\ f &\longmapsto \big(n \mapsto (s \mapsto f(s.n)) \big) \end{aligned}$$

and

$$\begin{aligned} \mathrm{Hom}_S(N, \mathrm{Hom}_R(S, M)) &\longrightarrow \mathrm{Hom}_R(\phi^*(N), M)\,, \\ g &\longmapsto \big(n \mapsto g(n)(1_S) \big)\,. \end{aligned}$$

It is not hard to check that these maps are mutually inverse.

EXERCISE 1.134. In general, the left and right adjoint of a functor are different, even if both of them exist. However, for the case of group algebras of finite groups, i.e., for $R = \mathbb{K}[H]$ and $S = \mathbb{K}[G]$ with some group homomorphism $\iota\colon H \to G$, the map

$$\begin{aligned} \mathrm{Hom}_R(S, M) &\longrightarrow S \otimes_R M\,, \\ \Phi &\longmapsto \textstyle\sum_{g \in G} g \otimes \Phi(g) \end{aligned}$$

gives a natural transformation $\iota^* \to \iota_!$ from coinduction to induction.

Show that if the field $\mathbb{K}$ has characteristic zero, then this natural transformation is an isomorphism, so that we deal with an ambidextrous adjunction.

Consider the special case $G = \mathbb{Z}_2 = H$ with $\iota = \mathrm{id}\colon \mathbb{Z}_2 \to \mathbb{Z}_2$ and $\mathbb{K}$ a field of characteristic 2 to obtain an example in which the natural transformation $\iota^* \to \iota_!$ from coinduction to induction is actually zero.

EXERCISE 1.135. Let R and S be two rings, and let X be an (S, R)-bimodule. Prove that the functor $X \otimes_R -\colon R\text{-mod} \to S\text{-mod}$ is left adjoint to the functor $\mathrm{Hom}_S(X, -)\colon S\text{-mod} \to R\text{-mod}$.

This adjunction is often referred to as the *tensor-Hom adjunction*.

The tensor-Hom adjunction relates to Example 1.133 as follows: Any morphism $\phi\colon R \to S$ of rings makes S an (S, R)-bimodule by the prescription $s.s'.r := ss'\phi(r)$ for all $r \in R$ and all $s, s' \in S$. We denote this bimodule by S_ϕ. Then the functor $\phi_!$ from Example 1.133 coincides with $S_\phi \otimes_R -$.

A crucial feature of left and right adjoint functors is the preservation of colimits and limits, respectively. In fact, very often this is one of the main motivation for trying to find a pair of adjoint functors. In order to state this important result, we need to formalize what it means for a functor to preserve (co)limits.

DEFINITION 1.136. A functor $F\colon \mathcal{C} \to \mathcal{D}$ is called *continuous* if it preserves limits, in the sense that for any limit Y of a diagram X in $\mathcal{C}$, as exhibited by a cone $\alpha\colon Y \to X$, the map $F(\alpha)\colon F(Y) \to F(X)$ exhibits $F(Y)$ as a limit of the diagram $F(X)$ in $\mathcal{D}$.

Dually, a functor preserving colimits is called *cocontinuous*.

Proposition 1.101 implies that a functor is already continuous if it preserves products and equalizers.

EXERCISE 1.137. Prove that for any object X in a category $\mathcal{C}$ the Hom functor $\mathrm{Hom}(X,-)\colon \mathcal{C} \to \mathcal{S}et$ is continuous.

EXERCISE 1.138. Prove that any equivalence between categories is continuous and cocontinuous.

THEOREM 1.139. *Every left adjoint functor is cocontinuous. Every right adjoint functor is continuous.*

PROOF. Let $F\colon \mathcal{C} \to \mathcal{D}$ be a left adjoint to $G\colon \mathcal{D} \to \mathcal{C}$. Let X be a diagram in $\mathcal{C}$ of shape $\mathcal{I}$ such that $\alpha\colon X \to Y$ exhibits $Y \in \mathcal{C}$ as a colimit of X. We need to show that $F(\alpha)\colon F(X) \to F(Y)$ is a universal cocone to $F(X)$. To see this, consider an arbitrary cocone $\beta\colon F(X) \to Z$ to $F(X)$. The components $\beta_i\colon F(X(i)) \to Z$ correspond bijectively to morphisms $\beta_i'\colon X(i) \to G(Z)$, since the adjunction $F \dashv G$ provides us with natural isomorphisms $\mathrm{Hom}_{\mathcal{D}}(F(X(i)), Z) \cong \mathrm{Hom}_{\mathcal{C}}(X(i), G(Z))$. By naturality of these isomorphisms, the morphisms β_i' form a cocone $\beta'\colon X \to G(Z)$ to X. Further, by universality of the cocone $\alpha\colon X \to Y$ there is a unique morphism $\varphi\colon Y \to G(Z)$ such that $\varphi \circ \alpha = \beta'$. Again via the adjunction $F \dashv G$, the morphism $\varphi\colon Y \to G(Z)$ corresponds to a morphism $\varphi'\colon F(Y) \to Z$. This morphism is the unique morphism satisfying $F(\alpha)\varphi' = \beta$.

The proves that $F(\alpha)\colon F(X) \to F(Y)$ is universal, and hence that F preserves colimits. The statement for right adjoints is obtained by dualization. □

EXERCISE 1.140. Use Theorem 1.139 to prove that the functor F from $\mathcal{S}et$ to $\mathcal{V}ect(\mathbb{K})$ that sends a set X to the free vector space $F(X)$ generated by X sends disjoint unions to direct sums. That is, show that for any family $(X_j)_{j\in J}$ of sets there is a natural isomorphism $F(\bigsqcup_{j\in J} X_j) \cong \bigoplus_{j\in J} F(X_j)$.

EXERCISE 1.141. Show that right adjoint functors preserve monomorphisms, while left adjoint functors preserve epimorphisms.

Hint: For the former statement, first use the universal property of pullbacks to prove that a morphism $f\colon X \to Y$ in a category $\mathcal{C}$ is a monomorphism if and only if the square

$$\begin{array}{ccc} X & \xrightarrow{\mathrm{id}_X} & X \\ {\scriptstyle \mathrm{id}_X}\downarrow & & \downarrow{\scriptstyle f} \\ X & \xrightarrow[f]{} & Y \end{array}$$

is a pullback. Then invoke Theorem 1.139.

EXAMPLE 1.142. In Exercise 1.135 we established that for two rings R and S and an (S,R)-bimodule X, the functor $X \otimes_R -$ from R-mod to S-mod is left adjoint to the functor $\mathrm{Hom}_S(X,-)\colon S\text{-mod} \to R\text{-mod}$. By means of Theorem 1.139 this provides us, without any computation, with canonical isomorphisms

$$X \otimes_R \Big(\bigoplus_{j\in J} Y_j\Big) \cong \bigoplus_{j\in J} (X \otimes_R Y_j)$$

for any family $(Y_j)_{j\in J}$ of R-modules, and

$$\mathrm{Hom}_S\Big(X, \prod_{j\in J} Z_j\Big) \cong \prod_{j\in J} \mathrm{Hom}_S(X, Z_j)$$

for any family $(Z_j)_{j\in J}$ of S-modules.

Consider now a short exact sequence

$$0 \longrightarrow A \xrightarrow{\iota} B \xrightarrow{\pi} C \longrightarrow 0$$

of R-modules. Recall that this means that at each term in the sequence the image of the incoming map coincides with the kernel of the outgoing map. Explicitly, we have:

(1) ι is injective;

(2) π is surjective;

(3) the image of ι is the kernel of π (implying in particular $\pi \circ \iota = 0$).

Now the sequence

$$X \otimes_R A \longrightarrow X \otimes_R B \longrightarrow X \otimes_R C \longrightarrow 0 \tag{1.143}$$

induced by functoriality of $X \otimes_R -$ is exact. (We express this by saying that $X \otimes_R -$ is a *right exact functor.*)

To prove this claim we observe that the points (2) and (3) above can be rephrased as the statement that C is the coequalizer of

$$A \underset{0}{\overset{\iota}{\rightrightarrows}} B\,.$$

It follows that the sequence (1.143) is exact because $X \otimes_R -$, being a left adjoint, preserves colimits.

Dually, $\mathrm{Hom}_S(X, -)$ sends a short exact sequence

$$0 \longrightarrow K \longrightarrow L \longrightarrow M \longrightarrow 0$$

of S-modules to an exact sequence

$$0 \longrightarrow \mathrm{Hom}_S(X, K) \longrightarrow \mathrm{Hom}_S(X, L) \longrightarrow \mathrm{Hom}_S(X, M)\,,$$

i.e., $\mathrm{Hom}_S(X, -)$ is a *left exact* functor.

Generally, $X \otimes_R -$ will fail to be left exact, and $\mathrm{Hom}_S(X, -)$ will fail to be right exact. This failure naturally leads to *left derived tensor products* and *right derived Hom spaces* in homological algebra, also known as *Tor* and *Ext*. We refer to Chapters 2 and 3 of [**Wei**] for details.

CHAPTER 2

Tensor products, duals, and braidings

According to Example 1.58, a tentative definition of a topological field theory is as a representation of a cobordism category $\mathcal{C}ob$, that is, as a functor from $\mathcal{C}ob$ to the category $\mathcal{V}ect$ of vector spaces. In the present chapter we start by outlining, in Section 2.1, an approach to a specific class of quantum field theories which can be interpreted as three-dimensional topological field theories – provided that a few heuristic ideas about quantum field theory of which the approach makes much use are properly translated into precise mathematical statements. Our attempt to achieve such a translation will in particular unveil the limitations of the tentative definition in Example 1.58. Specifically, we will see that it is far from sufficient to think of $\mathcal{C}ob$ and of $\mathcal{V}ect$ just as plain categories and of a topological field theory just as a plain functor between them. A necessary additional ingredient will be further structures both on categories and on functors. Sections 2.2 – 2.7 are devoted to introducing these further structures.

We are aware of the danger that, while the heuristic considerations in Section 2.1 may appeal to part of the readership, other readers might feel lost. In case this happens, the discouraged reader should not hesitate to disregard Section 2.1 entirely and jump forward to Section 2.2 at any time.

2.1. Beyond TFT as a functor – a motivation

In this Section we present a concrete class of quantum field theoretic models, standard features of which are the physical notions of *field configurations* and of *gauge symmetry*. In the first place, these notions arise at a heuristic level. Our aim in the sequel is to formulate these models in a precise manner so that they become amenable to rigorous mathematical reasoning. This goal will eventually be accomplished in Chapter 4. The way to this achievement will be paved by gathering various further categorical tools.

For concreteness, we will deal with models formulated in three dimensions. However, it is essential to take not only three-dimensional manifolds into consideration, but lower-dimensional ones as well.

We structure our discussion into a sequence of statements that can be regarded as (a toy version of) principles of quantum field theories. In this setting, $\mathbb{K}$ is taken to be the field $\mathbb{C}$ of complex numbers. We begin with the basic input data.

PRINCIPLES OF FIELD THEORY 2.1.

(1) Let G be a finite group, and let M be a compact oriented manifold of dimension 1, 2 or 3, possibly with boundary. A concrete model for the "*space*" $\mathcal{A}(M)$ *of field configurations* on M is provided by the category

$$\mathcal{A}(M) := \mathcal{B}un_G(M)$$

of principal G-bundles. In this context it is appropriate to think about a G-bundle as a solution to some "equation of motion". The existence of (iso)morphisms between bundles can be interpreted as indicating that the bundles overparametrize such solutions.

(2) One might thus be tempted to work instead with isomorphism classes of bundles. This would, however, preclude the possibility to build global solutions from local ones. Indeed, keeping a redundant parametrization of solutions together with the required identifications is indispensable for attaining *locality* in a gauge theory. The notion of locality is intimately linked with another aspect of bundles. Namely, a bundle on a manifold can be glued together from bundles on an open cover and transition functions on intersections of open sets in the cover. (In a special case, this will be stated in Proposition 4.64.) In this sense, bundles are local. Technically speaking, they satisfy *descent*. Locality is a crucial ingredient in many approaches to (quantum) field theory. A more precise discussion of the gluing of bundles and its relation with locality is beyond the level of this introductory text.

(3) The field configurations $\mathcal{B}un_G(M)$ form a *category* rather than a set – this is the reason why in (1) we have put the term "space" in quotation marks. In fact, as seen in Exercise 1.33, $\mathcal{B}un_G(M)$ is even a groupoid. The morphisms of the groupoid $\mathcal{B}un_G(M)$ account for the redundancy that is present when using bundles; in physics terminology, they constitute the *gauge transformations* of the field configurations. This way the language of categories allows us to systematically incorporate gauge transformations in quantum field theoretic models.

(4) A further piece of data needed to specify a model is a function

$$S(M): \quad \mathcal{A}(M) \longrightarrow \mathbb{C}$$

on the objects of the groupoid of field configurations. This function is the (topological) *action* of the model.

(5) According to the principle of equivalence which we formulated in Remark 1.61, the decisive aspects of a model should not change if we replace the groupoid of field configurations by an equivalent one.

We therefore require that the action functional S is *gauge invariant*, i.e., that it is constant on isomorphism classes in $\mathcal{A}(M)$. This requirement guarantees that $S(M)$ amounts to a function on the set of isomorphism classes of $\mathcal{A}(M)$, and hence makes sense on any equivalent model of $\mathcal{A}$ that we may decide to choose.

REMARK 2.2. The finite group G, i.e., the structure group of the principal bundles that we consider, can be regarded as a variant of a compact Lie group; we restrict our attention to finite groups in order to avoid technical difficulties. For a compact Lie group, the groupoid of G-bundles should be replaced by the groupoid $\mathcal{A}(M) := \mathcal{B}un_G^{\nabla}(M)$ of G-bundles with connection – or, in physics terminology: of G-bundles with a *gauge field*. (For a finite group, all connections are automatically flat, so this definition indeed extends the previous one for finite groups.) Analogously as in the case of a finite group, one can still specify a gauge invariant action functional $S\colon \mathcal{B}un_G^{\nabla}(M) \to \mathbb{C}$. If the action S can be written as an integral over M, then the integrand is called the *Lagrangian* (or Lagrangian density) of the model.

A specific type of action can often only be formulated for manifolds of a particular dimension. If the manifold M is three-dimensional, a valid possibility is to take the *Chern-Simons action*

$$S_{\mathrm{CS}}(M) = \frac{k}{4\pi} \int_M \mathrm{tr}\,(A \wedge \mathrm{d}A + \frac{2}{3}\, A \wedge A \wedge A)\,,$$

where A is a connection one-form on M that takes values in a simple finite-dimensional Lie algebra $\mathfrak{g}$ and k is a positive integer, called the level. For details, see, e.g., [**Wit**], [**Fr1**], and [**Koh**, Ch. 2.5].

On the other hand, actions that can be defined for manifolds of arbitrary dimension typically require further structure on the manifold, like a metric. A metric is, e.g., needed for formulating the gauge theories based on principal bundles with connection and with continuous structure group, like Yang-Mills theories or other gauge theories in particle physics; these differ in many respects from the finite gauge theories considered here.

We next analyze how, on the basis of physical principles, we can obtain mathematical structures of the type that appear in the tentative definition of a topological field theory in Example 1.58. The basic idea is to obtain an invariant $Z(M)$ for a closed oriented three-dimensional manifold M – in physics terminology, a *partition function* – by integrating over all field configurations $\Phi \in \mathcal{A}(M)$:

$$Z(M) := \text{``} \int_{\mathcal{A}(M)} \mathrm{d}\Phi \;\, \mathrm{e}^{\mathrm{i}\, S[\Phi]} \text{''}. \tag{2.3}$$

This formula involves a *path integral* which, in general, only has a heuristic meaning. When trying to obtain an accurate replacement for the heuristic formula (2.3), it would be a bad idea to define a measure by simply counting field configurations. Rather, the measure must account for the fact that field configurations form a category and that equivalent categories need not have the same number of objects. (It is not hard to find two equivalent categories for one of which the number of objects is finite while for the other it is infinite.)

A reasonable measure should instead count isomorphism classes, possibly amended by some suitable weighting factor. As we will explain in Section 4.1, when dealing with bundles with finite structure group over compact manifolds, one can indeed specify a suitable counting measure, namely the *groupoid cardinality*. Hereby the formal path integral expression reduces to a well-defined finite sum; this is spelled out in Proposition 4.6.

To formulate our gauge theoretic models in terms of a cobordism category $\mathcal{C}ob_{d,d-1}$, as needed in order to fit with the considerations in Example 1.58, it is crucial to admit manifolds with nonempty boundary. The role of manifolds with boundary comprises the following issues:

PRINCIPLES OF FIELD THEORY 2.4.

(1) Let M be an oriented three-manifold with two-dimensional boundary $\Sigma := \partial M$. Select a boundary field configuration $\phi \in \mathcal{A}(\Sigma)$ and consider the space

$$\mathcal{A}(M,\phi) := \{\Phi \in \mathcal{A}(M) \mid \Phi|_{\Sigma} = \phi\} \tag{2.5}$$

of all fields Φ on M that restrict to the prescribed boundary value ϕ on the boundary. In the case of bundles, this is again a category, with morphisms given by the gauge transformations that are the identity on the boundary Σ.

This step uses the fact that the restriction of fields along the embedding $\Sigma \longrightarrow M$ furnishes a functor $\mathcal{A}(M) \longrightarrow \mathcal{A}(\Sigma)$. Note that the possibility to restrict – or, more generally, *pull back* – field configurations appears as a key feature here.

(Formalizing the behavior of such pull back functors of the groupoids of field configurations leads to the notion of a *stack*. Here we refrain, however, from using the language of stacks. For pertinent information, see, e.g., [**Hei**].)

(2) Once we have specified boundary values ϕ of the field, we can more seriously think about performing a path integral over the fields. Accordingly, we introduce, still at a heuristic level, the complex number

$$Z(M)_\phi := \text{“}\int_{\mathcal{A}(M,\phi)} \mathrm{d}\Phi \ \mathrm{e}^{\mathrm{i}\,S[\Phi]}\text{”}.$$

This way every three-manifold M with boundary Σ provides an assignment

$$\begin{aligned} \psi_M : \quad \mathcal{A}(\Sigma) &\longrightarrow \mathbb{C}, \\ \phi &\longmapsto Z(M)_\phi . \end{aligned}$$

(3) The *principle of gauge invariance* stipulates that the value of the assignment ψ_M should be constant on isomorphism classes in $\mathcal{A}(\Sigma)$. We are thus led to assign to a manifold Σ of codimension 1 in M the vector space $\mathcal{H}_\Sigma$ of functions on $\mathcal{A}(\Sigma)$ constant on isomorphism classes. $\mathcal{H}_\Sigma$ is called the *state space* of Σ, and its elements are referred to as *states* or (in particular, in the physics literature) *wave functions*. Thus every three-manifold M with boundary Σ specifies a wave function in the state space $\mathcal{H}_\Sigma$.

(4) In case Σ is a disjoint union of two-manifolds, $\Sigma = \Sigma_1 \sqcup \Sigma_2$, then we know from Exercise 1.64 that $\mathcal{A}(\Sigma) \simeq \mathcal{A}(\Sigma_1) \times \mathcal{A}(\Sigma_2)$ for the corresponding categories of bundles, and hence the set of isomorphism classes of $\mathcal{A}(\Sigma)$ is a Cartesian product

$$\pi_0(\mathcal{A}(\Sigma)) \cong \pi_0(\mathcal{A}(\Sigma_1)) \times \pi_0(\mathcal{A}(\Sigma_2))$$

of those of $\mathcal{A}(\Sigma_1)$ and $\mathcal{A}(\Sigma_2)$. Since the free vector space on a Cartesian product $X \times Y$ is the tensor product of the free vector spaces on X and Y, this implies that the state spaces assigned to the manifolds are related by

$$\mathcal{H}_\Sigma \cong \mathcal{H}_{\Sigma_1} \otimes \mathcal{H}_{\Sigma_2} . \tag{2.6}$$

We conclude that in a topological field theory the assignment of vector spaces to codimension-1 submanifolds should be compatible with the disjoint union and the tensor product.

(5) The situation naturally extends to morphisms in $\mathcal{C}ob_{3,2}$, i.e., to cobordisms $M\colon \Sigma \longrightarrow \Sigma'$: To a cobordism M we wish to associate a linear map

$$Z(M): \quad \mathcal{H}_\Sigma \longrightarrow \mathcal{H}_{\Sigma'} .$$

We describe the map $Z(M)$ through numbers $Z(M)_{\phi,\phi'}$ for any pair of prescribed boundary values $\phi \in \mathcal{A}(\Sigma)$ and $\phi' \in \mathcal{A}(\Sigma')$. Heuristically, the number $Z(M)_{\phi,\phi'}$ is given by a path integral

$$Z(M)_{\phi,\phi'} := \text{``}\int_{\mathcal{A}(M,\phi,\phi')} \mathrm{d}\Phi \ \mathrm{e}^{\mathrm{i}\, S[\Phi]} \text{''},$$

where $\mathcal{A}(M,\phi,\phi')$ is the space of field configurations on M that restrict to the field configuration ϕ on the incoming boundary Σ and to the field configuration ϕ' on the outgoing boundary Σ'. The numbers $Z(M)_{\phi,\phi'}$ are called *transition amplitudes*, or also the *matrix elements* of $Z(M)$.

(6) The linear maps $Z(M)$ must be compatible with gluing of cobordisms along boundaries. This requirement is implemented by stipulating that Z is a functor from $\mathcal{C}ob_{3,2}$ to the category of vector spaces.

(7) Admitting the presence of boundaries allows us to reverse the process of gluing cobordisms. Indeed, one fruitful technique in topology is to study aspects of a closed three-manifold M by cutting it up into several simpler three-manifolds with boundary.

In particular, obtaining the invariant for M is in this way reduced to computing the invariants for those simpler pieces. Put differently, information about a manifold M can be inferred from information about simple manifolds from which M can be obtained by gluing, and about the specific way of gluing.

REMARK 2.7. By gluing two handlebodies P and Q of the same genus by an orientation reversing homeomorphism f from the boundary of P to the boundary of Q one obtains a compact oriented three-manifold $M = P \cup_f Q$. This is called a *Heegard splitting* of M. Every closed orientable three-manifold can be obtained in this manner. A given three-manifold M can admit Heegard splittings with different values of the genus g; the minimal possible value of g is called the Heegard genus of M.

Heegard splittings are special cases of *handle decompositions*, which exist for manifolds of any dimension. A handle decomposition of a d-dimensional manifold is a finite ascending chain

$$\emptyset = M_{-1} \subset M_0 \subset M_1 \subset M_2 \subset \cdots \subset M_{m-1} \subset M_m = M$$

of d-manifolds (possibly with boundary) M_i such that each M_i is obtained from M_{i-1} by attaching one out of a set of standard manifolds with boundary, called handles. For details we refer to [**GomP, Mat**].

EXERCISE 2.8. Show that the three-sphere S^3 can be obtained by gluing handlebodies of genus 1, as well as by gluing handlebodies of genus 2.

REMARK 2.9. An obvious question is whether this process of decomposing manifolds can be iterated, i.e., whether a codimension-1 manifold can be chopped again into smaller pieces, thereby introducing codimension-2 manifolds into the picture. Including manifolds of codimension 1 amounts to a two-layered structure and thus leads naturally to categories. The next step requires a three-layered structure, and thus bicategories, for an appropriate discussion. This issue will be addressed in Chapter 5.

As we want to be able to consider the gluing of disjoint manifolds, the topological field theory Z must assign, in a controlled manner, quantities also to disjoint unions of manifolds. And as seen in formula (2.6), for a disjoint union of two-manifolds the relevant structure is the tensor product of vector spaces. Moreover, with respect to the tensor product, the ground field $\mathbb{K}$, as a vector space over itself, is distinguished by the fact that, canonically, $\mathbb{K}\otimes V \cong V$, i.e., $\mathbb{K}$ behaves as a unit for the tensor product. Altogether this suggests that our model should supply us with suitable vector spaces and be compatible with their tensor product and with the existence of a unit in an appropriate sense. This requirement naturally leads to the notion of a monoidal category, which will be introduced in Section 2.2.

EXAMPLE 2.10. For any finite group G an example of a d-dimensional topological field theory is obtained by taking the simplest choice of an action one may think of, namely $S[P] := 0$ for every G-bundle $P \in \mathcal{A}(M) = \mathcal{B}un_G^{\nabla}(M)$.

There is also a richer variant, which makes in addition use of *group cohomology*, and which constitutes a useful toy example for studying topological terms in quantum field theory. (Readers unfamiliar with this notion may wish to skip the following example in a first reading; for an introduction to group cohomology see, e.g., Chapter 6 of [**Wei**].)

EXAMPLE 2.11. Consider as a Lagrangian a d-cocycle $\omega \in Z^d(G, \mathbb{K}^\times)$ in group cohomology with coefficients in the group $\mathbb{K}^\times$ of units of the field $\mathbb{K}$. The integration that yields the corresponding action S_ω can then be described as follows.

A crucial ingredient is the *classifying space* BG of the group G; this is, by definition, a connected CW-complex whose fundamental group is G and whose higher homotopy groups are all trivial. The space BG comes with a principal G-bundle $EG \longrightarrow BG$ with contractible total space EG that has the universal property that for any principal G-bundle P over some manifold M there is a map $f\colon M \longrightarrow BG$, uniquely determined up to homotopy, that satisfies $f^*EG \cong P$; see, e.g., Chapter 4.13 of [**Hus**]. One calls f a *classifying map* for the principal G-bundle P.

The d-cocycle ω on G can equivalently be described as a singular cocycle on the classifying space BG of G. For any classifying map $f\colon M \longrightarrow BG$ for a G-bundle P over M, the pullback $f^*\omega$ of ω along f is a d-cocycle on M. The action S_ω is then given by evaluation on the fundamental class $\mu_M \in Z_d(M;\mathbb{Z})$; this yields the number

$$S_\omega[P] := \langle f^*\omega, \mu_M \rangle \ \in \mathbb{K}^\times .$$

The cocycle ω may be regarded as a d-dimensional discrete variant of the Chern-Simons action that was mentioned in Remark 2.2; see [**FrQ**] for a more detailed discussion.

The groupoid $\mathcal{B}un_G(S^1)$ of G-bundles over the circle which, as explained in Example 1.74, is (noncanonically) equivalent to the groupoid $G/\!/G$, can also be described more invariantly as the fundamental groupoid of the space of continuous maps from the circle to the classifying space BG.

2.2. Monoidal categories

A major problem with our tentative definition of a topological field theory in Example 1.58 is that it neglects crucial structure that the two categories involved possess: the tensor product of vector spaces and the disjoint union of manifolds.

We improve the situation by taking this structure into account. To this end we would like to formalize the notion of tensor product in such a way that it makes sense also for more general categories $\mathcal{C}$ than $\mathcal{V}ect$. This is achieved by endowing $\mathcal{C}$ with the additional structure of a functor

$$\otimes : \quad \mathcal{C} \times \mathcal{C} \longrightarrow \mathcal{C} ,$$

from the Cartesian product of $\mathcal{C}$ with itself to $\mathcal{C}$, which is then called a tensor product functor, or simply a *tensor product*, on $\mathcal{C}$.

A tensor product functor associates to any pair (V, W) of objects an object $V \otimes W$, and to any pair (f, g) of morphisms a morphism $f \otimes g$ with source and target given by the tensor products of the source and target objects, i.e., $f \otimes g \colon V \otimes W \longrightarrow V' \otimes W'$ for $f \colon V \longrightarrow V'$ and $g \colon W \longrightarrow W'$.

Since $\otimes$ is by definition required to be a functor, we have

$$\mathrm{id}_{V \otimes W} = \mathrm{id}_V \otimes \mathrm{id}_W$$

as well as the *interchange law*

$$(f' \otimes g') \circ (f \otimes g) = (f' \circ f) \otimes (g' \circ g) \tag{2.12}$$

for any two pairs of composable morphisms.

In order to be deservedly called *tensor product*, the functor $\otimes$ should be demanded to be associative and unital. However, in line with the general principle that it is typically unwise to require equality of objects, as well as with the principle of equivalence, we should better not require associativity and unitality strictly, but rather only up to coherent natural isomorphism. In the case of vector spaces, there is indeed a reason why the two functors

$$\otimes \circ (\mathrm{id} \times \otimes) \quad \text{and} \quad \otimes \circ (\otimes \times \mathrm{id}) : \quad \mathcal{C} \times \mathcal{C} \times \mathcal{C} \longrightarrow \mathcal{C}$$

are isomorphic. Namely, for all vector spaces U, V, W, and X, each of the spaces $\mathrm{Hom}_{\mathcal{V}ect}((U \otimes V) \otimes W, X)$ and $\mathrm{Hom}_{\mathcal{V}ect}(U \otimes (V \otimes W), X)$ is in canonical bijection with the space of trilinear maps from the Cartesian product $U \times V \times W$ to X. These identifictions furnish a distinguished isomorphism

$$a : \quad \otimes \circ (\mathrm{id} \times \otimes) \longrightarrow \otimes \circ (\otimes \times \mathrm{id}) ,$$

which turns out to be natural. Analogously, also in the general case, instead of imposing a strict equality $(U \otimes V) \otimes W = U \otimes (V \otimes W)$ for all triples of objects U, V, and W in a category $\mathcal{C}$ with a tensor product functor $\otimes$, it is natural to require the existence of isomorphisms

$$a_{U,V,W} : \quad (U \otimes V) \otimes W \xrightarrow{\cong} U \otimes (V \otimes W)$$

(subject to certain natural consistency conditions) and take these isomorphisms as an additional part of the structure. The precise behavior of the natural isomorphisms $a_{U,V,W}$ is explained in the following

DEFINITION 2.13. A *monoidal category* consists of a category $(\mathcal{C}, \otimes)$ with tensor product, an object $\mathbf{1} \in \mathcal{C}$, called the *monoidal unit*, and natural isomorphisms

$$a: \quad \otimes \circ (\otimes \times \mathrm{id}) \xrightarrow{\cong} \otimes \circ (\mathrm{id} \times \otimes)$$

of functors $\mathcal{C} \times \mathcal{C} \times \mathcal{C} \to \mathcal{C}$ and

$$r: \quad \mathrm{id} \otimes \mathbf{1} \to \mathrm{id} \qquad \text{and} \qquad l: \quad \mathbf{1} \otimes \mathrm{id} \to \mathrm{id}$$

of functors $\mathcal{C} \to \mathcal{C}$ such that the following relations are satisfied:

- The *pentagon axiom*:
  For every quadruple of objects $U, V, W, X \in \mathrm{Obj}(\mathcal{C})$ the diagram

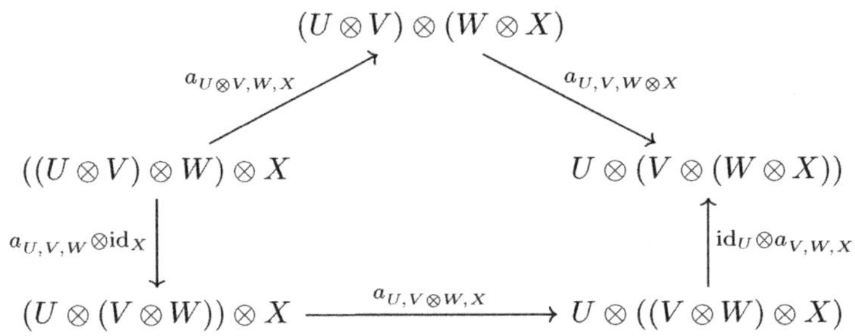

  commutes.
- The *triangle axiom*:
  For every pair of objects $V, W \in \mathrm{Obj}(\mathcal{C})$ the diagram

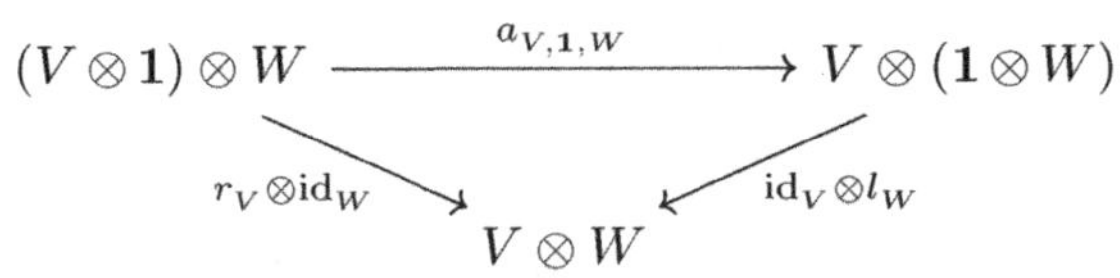

  commutes.

A monoidal category is called *strict* if the natural transformations a, l, and r are identities.

REMARKS 2.14.

(1) Instead of "monoidal category," also the term *tensor category* is in use. We avoid this terminology because in the literature it can instead also refer to monoidal categories with (varying types of) additional structure. The monoidal unit $\mathbf{1}$ is also called the *tensor unit*, or also the *unit object*.

(2) The natural isomorphism a is called the *associator*, or *associativity constraint*, of $\mathcal{C}$, and l and r are called the left and right *unitors*, or *unit constraints*, respectively.

(3) By the definition of a natural transformation, the associator a has as components for every triple U, V, W of objects an isomorphism

$$a_{U,V,W}: \quad (U \otimes V) \otimes W \longrightarrow U \otimes (V \otimes W)$$

such that all diagrams of the form

$$\begin{array}{ccc} (U \otimes V) \otimes W & \xrightarrow{a_{U,V,W}} & U \otimes (V \otimes W) \\ \Big\downarrow{\scriptstyle (f\otimes g)\otimes h} & & \Big\downarrow{\scriptstyle f\otimes(g\otimes h)} \\ (U' \otimes V') \otimes W' & \xrightarrow{a_{U',V',W'}} & U \otimes (V \otimes W) \end{array}$$

commute.

(4) A monoidal category can be regarded as a higher analogue of an associative, unital monoid, hence the qualification "monoidal". The *associator* a is, however, a *structure*, rather than a *property*. A property is imposed at the level of natural transformations, in the form of the pentagon axiom. For a given category $\mathcal{C}$ and a given tensor product $\otimes$, there can exist inequivalent associators; a simple illustration will be given in Example 2.47.

(5) The pentagon axiom guarantees that any two isomorphisms that are paths between two given objects in a formal diagram with morphisms built only from associators and unit constraints are equal or, in other words, that any change of the bracketing in a multiple tensor product is realized by a unique isomorphism built from associators and unitors. This result is known as *Mac Lane's coherence theorem*. We refer to [**Mac**, Ch. VII.2] for a detailed exposition, including the precise notion of formal diagram.

EXAMPLES 2.15.

(1) The category of vector spaces over a fixed field $\mathbb{K}$ is a monoidal category. It is not strict, however: the vector spaces $(U \otimes V) \otimes W$ and $U \otimes ((V \otimes W)$ are defined by different universal properties and are thus merely isomorphic rather than identical.

In linear algebra classes, it is common practice to tacitly replace the non-strict monoidal category of vector spaces by an equivalent strict monoidal category. This is justified by the strictification result described in Remark 2.46.

(2) Let $\mathcal{C}$ be a small category. The endofunctors

$$F: \quad \mathcal{C} \longrightarrow \mathcal{C}$$

are the objects of a strict monoidal category, denoted by $\mathcal{E}nd(\mathcal{C})$. The morphisms in $\mathcal{E}nd(\mathcal{C})$ are natural transformations, and the tensor product is composition of functors.

(3) There are two obvious monoidal structures on the category $\mathcal{S}et$ of sets. For the first, the tensor product is given by disjoint union and the monoidal unit is the empty set. For the second, the tensor product is given by the Cartesian product and the monoidal unit is the set with one element.

(4) For each $d \geq 1$, the category $\mathcal{C}ob_{d,d-1}$ can be endowed with the structure of a monoidal category. The tensor product

$$\otimes : \quad \mathcal{C}ob_{d,d-1} \times \mathcal{C}ob_{d,d-1} \longrightarrow \mathcal{C}ob_{d,d-1}$$

is given by disjoint union $\sqcup$. The unit object of $\mathcal{C}ob_{d,d-1}$ is the empty set $\emptyset$, regarded as a smooth manifold of dimension $d-1$.

(For the case $d=0$, i.e., the category $\mathcal{C}ob_{1,0}$, we have encountered the disjoint union already in our motivating Example 1.1.)

(5) For G a group, the category $\mathcal{V}ect_G(\mathbb{K})$ of G-graded $\mathbb{K}$-vector spaces described in Example 1.5(5), i.e., of $\mathbb{K}$-vector spaces with a direct sum decomposition

$$V = \bigoplus_{g \in G} V_g$$

is monoidal. The tensor product $V \otimes W$ in $\mathcal{V}ect_G(\mathbb{K})$ is G-bi-graded, i.e., decomposes as a direct sum as $V \otimes W = \bigoplus_{g,h \in G} V_g \otimes W_h$; it becomes G-graded by writing

$$V \otimes W = \bigoplus_{g \in G} \Big(\bigoplus_{h \in G} V_h \otimes W_{h^{-1}g} \Big) .$$

Together with the associativity constraint inherited from $\mathcal{V}ect(\mathbb{K})$, and with $\mathbb{K}_e$, i.e., the ground field $\mathbb{K}$ in homogeneous degree $e \in G$, as the tensor unit, $\mathcal{V}ect_G(\mathbb{K})$ is a monoidal category.

(6) For the considerations in (5), inverses in G are not needed. Thus we could also consider vector spaces graded by any unital associative monoid.

(7) Let A be an algebra over a field $\mathbb{K}$. The category A-$\mathcal{B}imod$ of A-bimodules has A-bimodules as objects and bimodule maps as morphisms. A monoidal structure on A-$\mathcal{B}imod$ is given by the relative tensor product over A. In more detail, in the monoidal product $M \otimes_A N$ of two A-bimodules M and N the right A-action on M and the left A-action on N are used to form the relative tensor product, while the left A-action on M and the right A-action on N endow $M \otimes_A N$ with the structure of an A-bimodule. The monoidal unit of A-$\mathcal{B}imod$ is the *regular* A-bimodule, i.e., the vector space A with both left and right action given by the multiplication.

REMARK 2.16. There is no tensor product of infinite-dimensional Hilbert spaces. More precisely, the category of Hilbert spaces and continuous linear maps does not admit a tensor product. Indeed, while setting $\langle v \otimes w, v' \otimes w' \rangle := \langle v, v' \rangle \, \langle w, w' \rangle$ endows the *algebraic* tensor product $V \otimes_{\text{alg.}} W$ of two Hilbert spaces V and W with an inner product and the completion of $V \otimes_{\text{alg.}} W$ with respect to the associated norm is again a Hilbert space, the so obtained Hilbert space does not possess the universal property of a tensor product of vector spaces. For more details see [**Gar**] and also Example 4.28 of [**Lim**].

GRAPHICAL DESCRIPTION 2.17. Any strict monoidal category admits a graphical calculus for its morphisms that extends the basic string diagrams that were presented in Graphical Description 1.16. For a nonstrict monoidal category $\mathcal{C}$, the graphical calculus becomes available after replacing $\mathcal{C}$ with an equivalent strict monoidal category, thereby invoking the strictification result to be stated in Remark 2.46. (Also, while strictness is assumed in the graphical description of any single morphism, for an *equality* between morphisms, the associator and unitors can be restored unambiguously from the equality between the respective string diagrams, so this information is not lost when expressing the equality diagrammatically.)

A salient feature of the graphical calculus for categories as given in Graphical Description 1.16 is that it is intrinsically *one-dimensional*: owing to associativity of the composition, what matters is only the linear order of coupons that are inserted along a string. To be able to account for the tensor product, we need to draw the diagrams instead on a "canvas" that is *two-dimensional* or, more concretely, on some region in the plane, say a rectangle, as indicated by the dashed lines in the following picture:

(2.18)

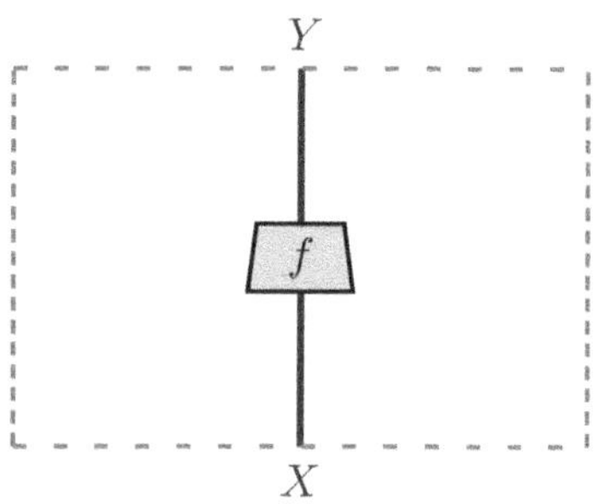

For all the pictures below, the canvas will be such a rectangle, but it will be suppressed throughout. We continue to allow for the local move (1.18), in which vertically adjacent coupons in a string diagram are concatenated. Just like that move is made possible by the existence of a composition map, the existence of the tensor product, together with the presence of a second dimension, now allows us to define another local move in which a pair of horizontally juxtaposed coupons is combined to form a single coupon:

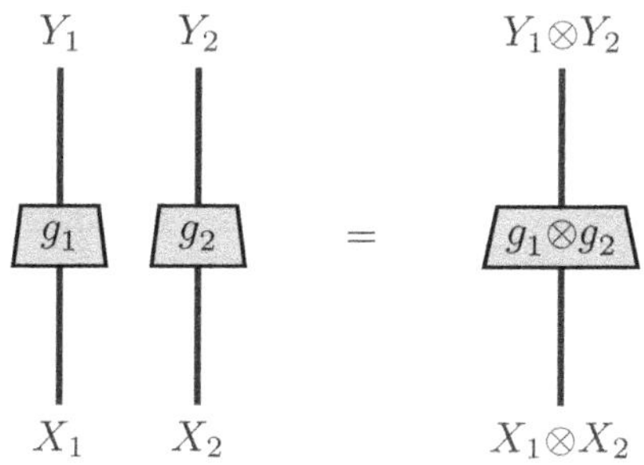

Also, indicating the identity endomorphism of the monoidal unit **1** by a dashed line, strictness of **1** amounts to the equalities

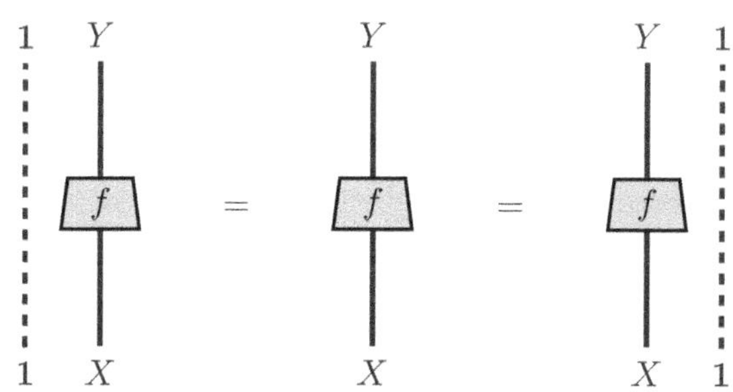

Accordingly, omitting a line labeled by the monoidal unit from a string diagram does not alter its meaning. This is sometimes expressed by saying that in a string diagram the monoidal unit is *transparent*, or also *invisible.*

It is often convenient to depict strings as curved rather than straight lines. When doing so, one implicitly agrees on the rule that the categorical interpretation of a string diagram remains unchanged under progressive planar isotopy. That is, strings can be smoothly deformed arbitrarily, as long as – as indicated by the qualification *progressive* – when proceeding from bottom to top (and, technically, after shrinking each coupon to a point), the tangent vector to a string always has a positive upward component. (Using the concept of an upward component, we implicitly regard the canvas, i.e., the rectangle in (2.18), as being 2-framed.) String diagrams that are related this way are to be considered as identical; as an illustration, we identify

(2.19)

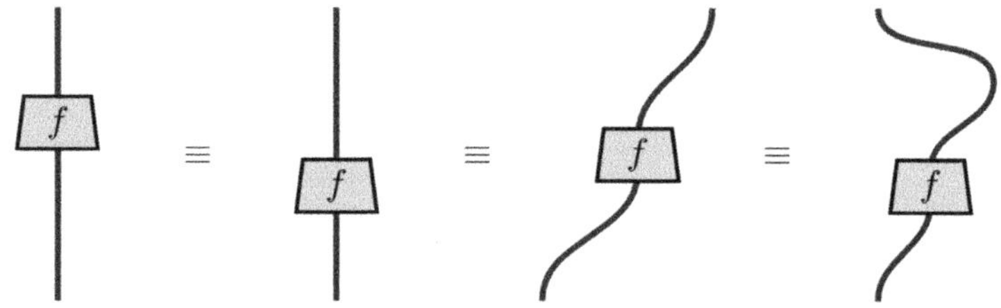

A precise formal statement of these identifications is given, e.g., in Theorem 1.2 of [**JS1**] and Theorem 3 of [**Sel**]. As one consequence of this rule, we note that the string diagram description of the interchange law (2.12) looks like:

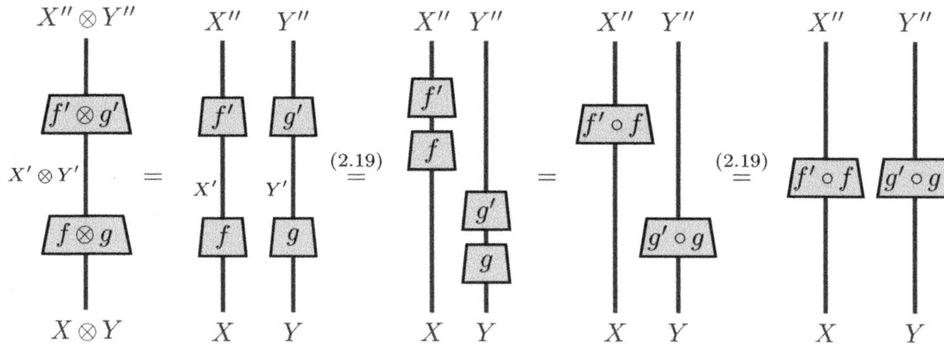

2.3. Bialgebras

Recall that an associative unital algebra over a field $\mathbb{K}$ is a triple (A, μ, η) consisting of a vector space A, an associative multiplication map $\mu\colon A \otimes A \longrightarrow A$ and a unit map $\eta\colon \mathbb{K} \longrightarrow A$. Also recall, from Definition 1.15, the example of a group algebra. Group algebras admit also further structure, which can be formalized as follows.

DEFINITION 2.20.

(1) Let $\mathbb{K}$ be a field. A (counital, coassociative) *coalgebra* over $\mathbb{K}$ is a triple (C, Δ, ε), consisting of a $\mathbb{K}$-vector space C and two $\mathbb{K}$-linear maps, the *coproduct* (or *comultiplication*)

$$\Delta : \quad C \longrightarrow C \otimes C$$

(with $\otimes \equiv \otimes_{\mathbb{K}}$ the tensor product of $\mathbb{K}$-vector spaces) and the *counit*

$$\varepsilon : \quad C \longrightarrow \mathbb{K},$$

that satisfy

$$(\Delta \otimes \mathrm{id}_C) \circ \Delta = (\mathrm{id}_C \otimes \Delta) \circ \Delta$$

(*coassociativity*) and

$$(\varepsilon \otimes \mathrm{id}_C) \circ \Delta = \mathrm{id}_C = (\mathrm{id}_C \otimes \varepsilon) \circ \Delta$$

(*counitality*).

(2) Given a coalgebra (C, Δ, ε), the *coopposite coalgebra* C^{copp} is the coalgebra $(C, \Delta^{\mathrm{copp}}, \varepsilon)$ with coproduct $\Delta^{\mathrm{copp}} := \tau_{C,C} \circ \Delta$, where

$$\tau_{C,C} : \quad c \otimes c' \longmapsto c' \otimes c \tag{2.21}$$

is the *flip map*, i.e., the endomorphism of $C \otimes C$ that exchanges the two tensor factors.

A coalgebra is called *cocommutative* if $\Delta^{\mathrm{copp}} = \Delta$.

(3) A *coalgebra map* between coalgebras C and C' is a linear map

$$\varphi : \quad C \longrightarrow C'$$

between the underlying vector spaces such that the equalities

$$\Delta' \circ \varphi = (\varphi \otimes \varphi) \circ \Delta \qquad \text{and} \qquad \varepsilon' \circ \varphi = \varepsilon$$

hold.

GRAPHICAL DESCRIPTION 2.22. We depict the comultiplication graphically as

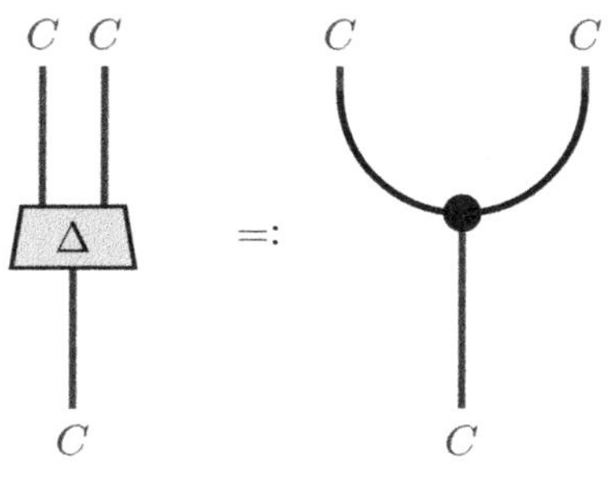

and the counit as

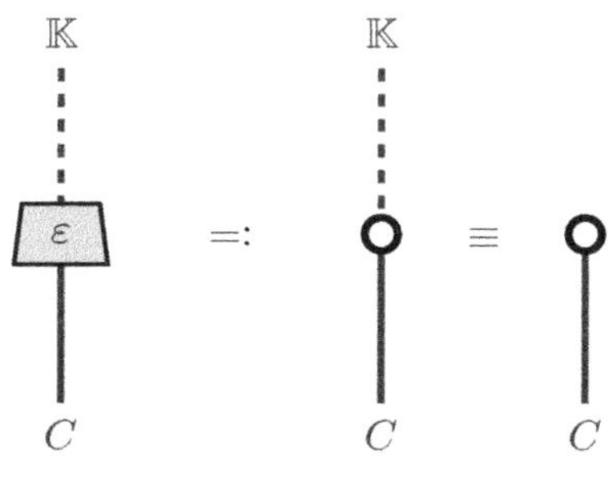

Then the coassociativity and the counit conditions read (omitting the obvious labels C, Δ and ε)

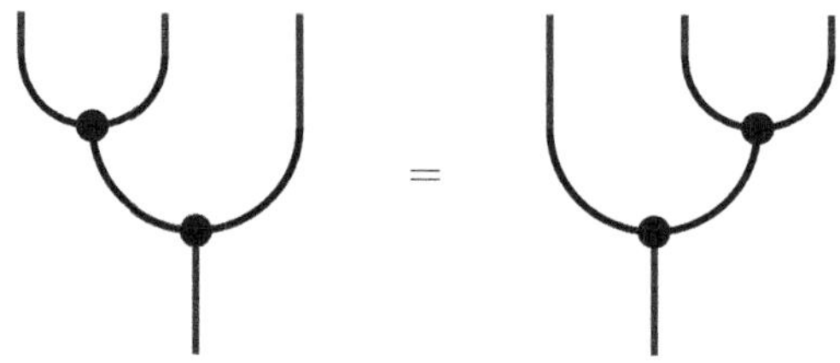

and

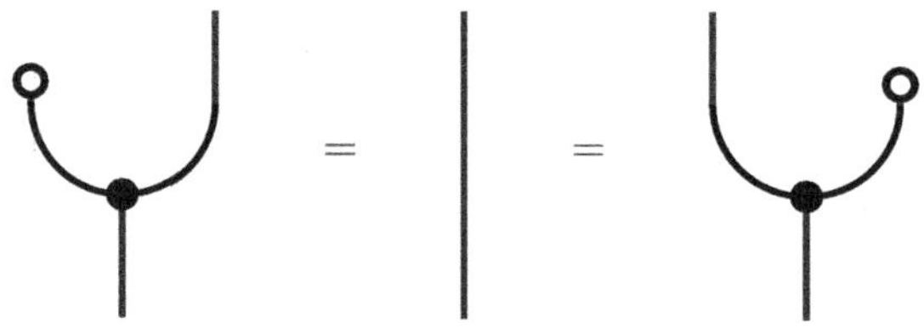

respectively. And the defining equalities of a coalgebra map $\varphi\colon C \longrightarrow C'$ look like

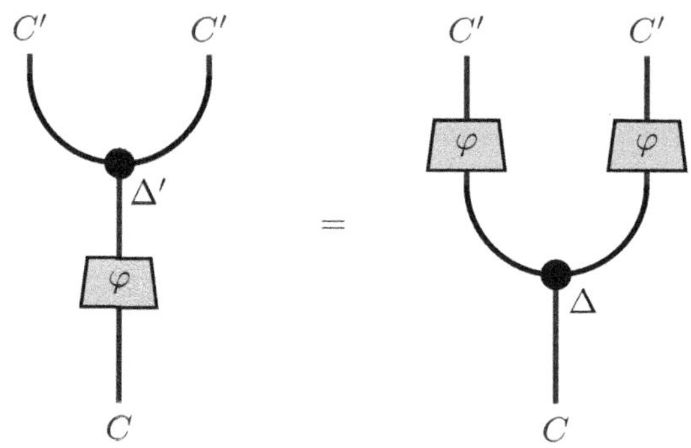

and

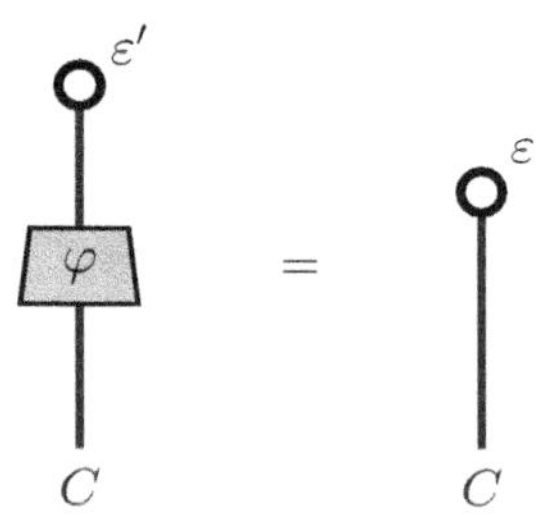

We can think of these pictures as string diagrams in the monoidal category of $\mathbb{K}$-vector spaces. Actually, as string diagrams they make sense for any other monoidal category as well, and indeed one can define algebras (A, μ, η) and coalgebras (C, Δ, ε) in any monoidal category, with the monoidal unit taking over the role of the field $\mathbb{K}$ in the unit and counit maps, and the categorical tensor product taking over the role of $\otimes_{\mathbb{K}}$ in the multiplication and comultiplication maps.

EXAMPLE 2.23. Let S be any set. Denote by $\mathbb{K}[S]$ the free $\mathbb{K}$-vector space with basis S. This vector space becomes a counital coassociative coalgebra with coproduct given by the diagonal map $\Delta(s) := s \otimes s$ and counit given by $\varepsilon(s) = 1$ for all $s \in S$. The so obtained coalgebra $\mathbb{K}[S]$ is cocommutative.

In particular, for any group G the group algebra $\mathbb{K}[G]$ is a cocommutative coalgebra.

EXAMPLE 2.24. A *Lie algebra* $\mathfrak{g}$ over a field $\mathbb{K}$ is a $\mathbb{K}$-vector space equipped with a bilinear map

$$\mathfrak{g} \times \mathfrak{g} \longrightarrow \mathfrak{g}, \quad (x, y) \longmapsto [x, y],$$

called the *Lie bracket*, that is antisymmetric, i.e., obeys $[x, y] = -[y, x]$, and satisfies the *Jacobi identity*

$$[x, [y, z]] + [y, [z, x]] + [z, [x, y]] = 0 .$$

A morphism of Lie algebras is a linear map respecting the Lie bracket.

For any associative $\mathbb{K}$-algebra A with product "$\cdot$" one obtains the structure of a Lie algebra $\mathbb{L}\mathrm{ie}(A)$ on the vector space underlying A by defining the Lie bracket to be given by the commutator, i.e.

$$[a, b] := a \cdot b - b \cdot a .$$

This provides a functor $\mathbb{L}\mathrm{ie}$ from the category of (unital) associative algebras to the category of Lie algebras. The functor $\mathbb{L}\mathrm{ie}$ admits a left adjoint $\mathbb{U}$ which assigns to a Lie algebra $\mathfrak{g}$ an associative unital algebra $\mathbb{U}(\mathfrak{g})$, called the *universal enveloping algebra* of $\mathfrak{g}$. By the definition of a left adjoint, morphisms from $\mathbb{U}(\mathfrak{g})$ into any associative algebra A are given by morphisms $\mathfrak{g} \longrightarrow \mathbb{L}\mathrm{ie}(A)$ of Lie algebras.
One can thus define a comultiplication on the algebra $\mathbb{U}(\mathfrak{g})$ via the morphism $\mathfrak{g} \longrightarrow \mathbb{L}\mathrm{ie}(\mathbb{U}(\mathfrak{g}) \otimes \mathbb{U}(\mathfrak{g}))$ of Lie algebras that is given by $x \longmapsto x \otimes 1 + 1 \otimes x$. (The reader should check that this is indeed a morphism of Lie algebras.) A counit for this coproduct is obtained by the algebra morphism $\mathbb{U}(\mathfrak{g}) \longrightarrow \mathbb{K}$ that is given by the Lie algebra map $x \longmapsto 0$. This prescription endows the universal envoloping algebra of any Lie algebra with the structure of a cocommutative coasssociative coalgebra.

EXAMPLE 2.25. An algebra in the strict monoidal category $\mathcal{E}nd(\mathcal{C})$ of endofunctors of a category $\mathcal{C}$ (see Examples 2.15(2)), with composition of functors as the monoidal product, is called a *monad* on $\mathcal{C}$. Every adjunction of functors gives rise to a monad. Indeed, with the terminology and notation for adjunctions introduced in Definition 1.127, the composite $A := G \circ F$ of the functors $F\colon \mathcal{C} \longrightarrow \mathcal{D}$ and $G\colon \mathcal{D} \longrightarrow \mathcal{C}$ is an endofunctor of $\mathcal{C}$, while the unit η of the adjunction gives a morphism from the identity endofunctor, i.e., the monoidal unit of $\mathcal{E}nd(\mathcal{C})$, to the object A. Further, A is equipped with a multiplication $A \circ A = G \circ F \circ G \circ F \longrightarrow G \circ F$ by using for the horizontal composition the counit $\varepsilon\colon F \circ G \longrightarrow \mathrm{id}_{\mathcal{D}}$ of the adjunction. One can show that every monad can be obtained from an adjunction, albeit the adjunction is typically not uniquely determined by the monad.

For further information on monads see, e.g., Chapter 5 of [**Ri2**] and Section 20 of [**AdHS**]. It is also worth mentioning that in part of the literature, modules over a monad T are called *algebras* over T.

EXERCISE 2.26. Let $\mathbb{K}$ be a field and A a be $\mathbb{K}$-algebra. Show that the forgetful functor from the category A-mod of A-modules to vector spaces and its left adjoint on the one hand, and the forgetful functor from the category $A\text{-mod}^{\text{free}}$ of *free* (or induced) A-modules to vector spaces and its left adjoint on the other, induce the same monad on the category $\mathcal{V}ect(\mathbb{K})$.

REMARKS 2.27.

(1) The counit is uniquely determined (if it exists).

(2) The following shorthand notation, due to Heyneman and Sweedler, is frequently called *Sweedler notation*. Let (C, Δ, ε) be a coalgebra. For any element $x \in C$ one can find finitely many elements $x_i' \in C$ and $x_i'' \in C$ such that

$$\Delta(x) = \sum_i x_i' \otimes x_i''.$$

We rewrite this expression by dropping summation indices and even omit the sum, according to

$$\Delta(x) = \sum_{(x)} x_{(1)} \otimes x_{(2)} \equiv x_{(1)} \otimes x_{(2)}.$$

In Sweedler notation, the counitality axiom reads

$$\varepsilon(x_{(1)})\, x_{(2)} = x = \varepsilon(x_{(2)})\, x_{(1)}$$

for all $x \in C$, while cocommutativity becomes

$$x_{(1)} \otimes x_{(2)} = x_{(2)} \otimes x_{(1)}$$

for all $x \in C$.

Finally, coassociativity reads

$$(x_{(1)})_{(1)} \otimes (x_{(1)})_{(2)} \otimes x_{(2)} = x_{(1)} \otimes (x_{(2)})_{(1)} \otimes (x_{(2)})_{(2)}$$

for all $x \in C$; we abbreviate this element also as $x_{(1)} \otimes x_{(2)} \otimes x_{(3)}$.

Given a $\mathbb{K}$-algebra (A, μ, η) with product

$$\mu: \quad A \otimes A \longrightarrow A, \qquad (a, b) \longmapsto a * b$$

and unit $\eta: \mathbb{K} \longrightarrow A$, the tensor product vector space $A \otimes A$ is naturally an algebra, with product

$$(a \otimes b) * (c \otimes d) := (a * c) \otimes (b * d)$$

for $a, b, c, d \in A$, and with unit $\eta \otimes \eta: \mathbb{K} \equiv \mathbb{K} \otimes \mathbb{K} \longrightarrow A \otimes A$.

From now on we equip $A \otimes A$ with this algebra structure, unless stated otherwise. Also, we regard $\mathbb{K}$ as an algebra, by multiplication of scalars.

DEFINITION 2.28. A *bialgebra* A over a field $\mathbb{K}$ is an algebra (A, μ, η) over $\mathbb{K}$ that also comes equipped with the structure of a coalgebra, in such a way that the coproduct $\Delta: A \longrightarrow A \otimes A$ and the counit $\varepsilon: A \longrightarrow \mathbb{K}$ are morphisms of algebras, that is,

$$\Delta \circ \mu = (\mu \otimes \mu) \circ (\mathrm{id}_A \otimes \tau_{A,A} \otimes \mathrm{id}_A) \circ (\Delta \otimes \Delta)$$

and

$$\varepsilon \circ \mu = \varepsilon \otimes \varepsilon,$$

as well as

$$\Delta \circ \eta = \eta \otimes \eta \qquad \text{and} \qquad \varepsilon \circ \eta = \mathrm{id}_{\mathbb{K}}.$$

REMARK 2.29. Equivalently, a bialgebra is a coalgebra equipped with the additional structure of an algebra such that the product and unit are morphisms of coalgebras.

Graphical Description 2.30. We depict the comultiplication and counit as in Graphical Description 2.22 and analogously the multiplication and unit of the algebra structure of H by

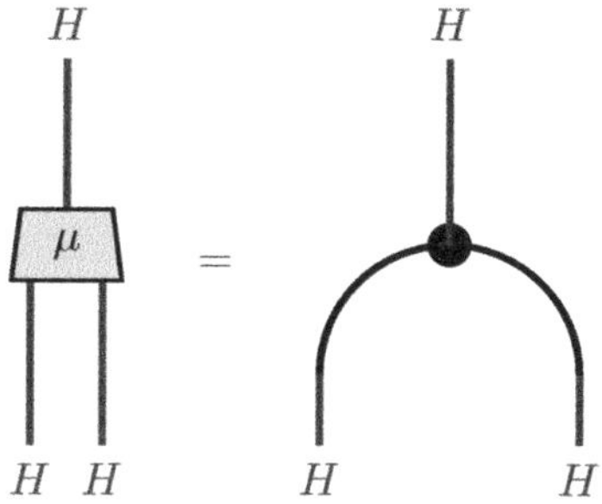

and

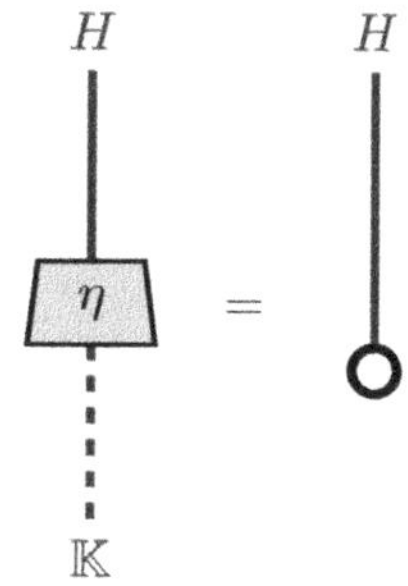

Then the string diagrams for the bialgebra axioms are as follows. The compatibility requirement between multiplication and comultiplication reads (omitting again all obvious labels)

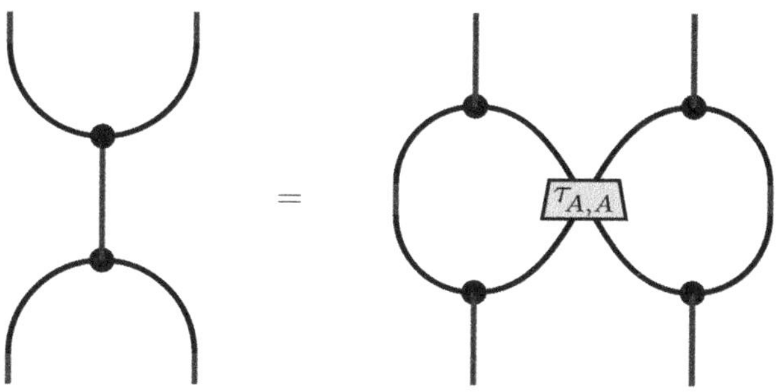

Those involving the unit and/or counit read

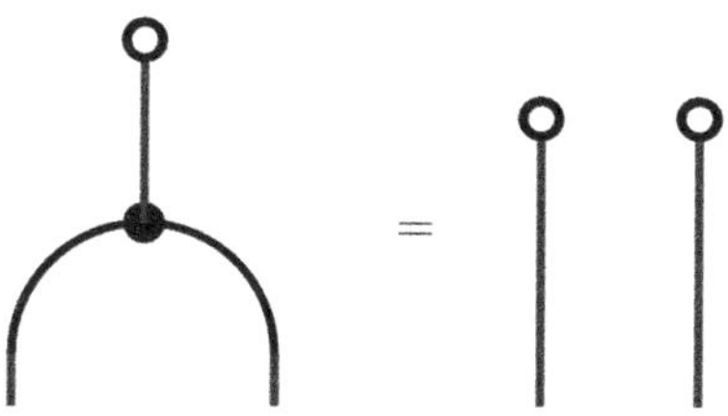

and

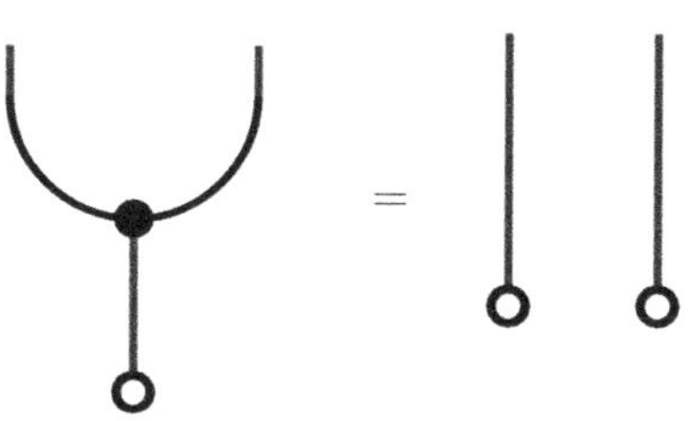

and

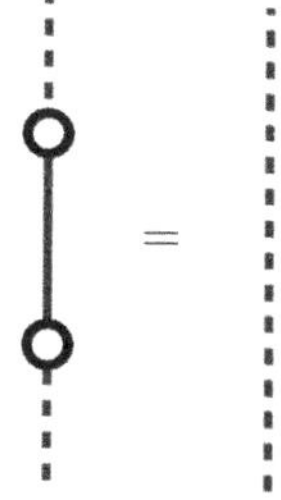

Again we can regard these pictures as string diagrams in the monoidal category of $\mathbb{K}$-vector spaces. In contrast, for a generic monoidal category there is no analogue of the flip map (2.21). Accordingly, in a generic monoidal category one can*not* define the notion of a bialgebra. Rather, this is only possible in monoidal categories that are endowed with a braiding, an additional structure to be introduced in Section 2.7.

EXERCISE 2.31. Show that for any group G, the group algebra $\mathbb{K}[G]$ with its coalgebra structure from Example 2.23 is a bialgebra.

EXERCISE 2.32 (Taft algebra). Let ζ be a complex Nth root of unity with $N \geq 2$. Consider the N^2-dimensional algebra H_N over $\mathbb{C}$ that is generated by two elements g and x subject to the relations

$$g^N = 1\,, \qquad x^N = 1 \qquad \text{and} \qquad xg = \zeta gx\,.$$

Show that by setting

$$\Delta(g) := g \otimes g\,, \qquad \Delta(x) := 1 \otimes x + x \otimes g$$

and

$$\varepsilon(g) := 1\,, \qquad \varepsilon(x) := 0$$

one obtains well-defined algebra morphisms $\Delta\colon H_N \longrightarrow H_N \otimes H_N$ and $\varepsilon\colon H_N \longrightarrow \mathbb{C}$, and that these maps endow H_N with the structure of bialgebra.

THEOREM 2.33. Let A be a bialgebra over a field $\mathbb{K}$.

(1) Let V and W be A-modules, with underlying vector spaces $\dot{V}$ and $\dot{W}$. Then by setting

$$a.(v \otimes w) := a_{(1)}.v \otimes a_{(2)}.w \tag{2.34}$$

for all $a \in A$, $v \in V$ and $w \in W$, the vector space $\dot{V} \otimes \dot{W}$ is endowed with the structure of an A-module, to be denoted by $V \otimes W$.

(2) The field $\mathbb{K}$ becomes an A-module by setting

$$a.\xi := \varepsilon(a)\, \xi$$

for $a \in A$ and $\xi \in \mathbb{K}$.

(3) These prescriptions yield a functor

$$\otimes : \quad A\text{-mod} \times A\text{-mod} \longrightarrow A\text{-mod}$$

which, together with the standard associators and unitors of the monoidal category of $\mathbb{K}$-vector spaces, extends to a monoidal structure on A-mod.

PROOF. For any algebra A, the vector space $V \otimes W$ becomes an $A{\otimes}A$-module via the prescription $(a \otimes b).(v \otimes w) := a.v \otimes b.w$ for $a, b \in A$, $v \in V$ and $w \in W$. Then $V \otimes W$, with the A-action (2.34), is the restriction of scalars (see Example 1.42(6)) of $V \otimes W$ as an $A{\otimes}A$-module, along the coproduct $\Delta\colon A \longrightarrow A \otimes A$ (that we can actually restrict along Δ uses that Δ is a morphism of algebras). This proves that $V \otimes W$, with the stated A-action, is indeed an A-module. It follows that the functor $\otimes\colon A\text{-mod} \times A\text{-mod} \longrightarrow A\text{-mod}$ is well-defined on objects. It can be extended to morphisms in a straightforward manner. In order to show that $\otimes$, together with the standard associators and unitors of the monoidal category of $\mathbb{K}$-vector spaces, forms a monoidal category, we need to prove that these associators and unitors are in fact morphisms of A-modules. This can be done by a direct computation. For example, the fact that for A-modules U, V and W the associator

$$a_{U,V,W} : \quad (U \otimes V) \otimes W \longrightarrow U \otimes (V \otimes W)$$

of U, V, and W, as $\mathbb{K}$-vector spaces, is actually a morphism of A-modules, follows from the coassociativity of the coproduct. We leave the details of this verification to the reader. □

EXAMPLE 2.35. If in Theorem 2.33 we take the bialgebra to be $A = \mathbb{K}[G]$ for a group G (see Example 2.31), then the monoidal product on the category of $\mathbb{K}[G]$-modules or, equivalently, G-representations, sends two G-representations V and W to the vector space $V \otimes W$ with G-action $g.(v \otimes w) = g.v \otimes g.w$ for $g \in G$, $v \in V$ and $w \in W$.

The notion of a monoidal category may be regarded as a *categorification* of the one of a monoid M, with the role of the elements of the monoid being taken over by the objects in the category, and the role of the multiplication taken over by the tensor product functor. In the same spirit as for rings, one can define the notion of a module over a monoid: a module over M is a set S together with a map $\rho\colon M \times S \longrightarrow S$, sending $(m, s) \mapsto s.m$, such that $(s_1 \cdot s_2).m = s_1.(s_2.m)$. Proceeding in the same vein for categories, one arrives at the structure of a *module category.*

Under categorification, properties tend to turn into structure. Specifically, while the multiplication in a ring is (strictly) associative, the tensor product in a monoidal category is associative only up to coherent isomorphisms, encoded in the structure of an associator. Analogously, in the case of module categories the representation property of the action of a ring on a module turns to an action of the monoidal category that satisfies the representation property only up to coherent isomorphisms.

DEFINITION 2.36. A (left) *module category* over a monoidal category

$$(\mathcal{C}, \otimes, \mathbf{1}, a, r, l)$$

consists of a category $\mathcal{M}$, an *action functor*

$$\triangleright\colon \quad \mathcal{C} \times \mathcal{M} \longrightarrow \mathcal{M},$$

and natural isomorphisms

$$a^{\triangleright} : \quad \triangleright \circ (\otimes \times \mathrm{id}) \xrightarrow{\cong} \triangleright \circ (\mathrm{id} \times \triangleright)$$

of functors $\mathcal{C} \times \mathcal{C} \times \mathcal{M} \to \mathcal{M}$ and

$$\lambda : \quad \mathbf{1} \triangleright \mathrm{id} \xrightarrow{\cong} \mathrm{id}$$

of functors $\mathcal{M} \to \mathcal{M}$ that are subject to the pentagon and triangle axioms

$$\begin{tikzcd}[column sep=small]
& (U \otimes V) \otimes (W \triangleright M) \arrow[dr, "a^{\triangleright}_{U,V,W\triangleright M}"] & \\
((U \otimes V) \otimes W) \triangleright M \arrow[ur, "a^{\triangleright}_{U\otimes V,W,M}"] \arrow[d, "a_{U,V,W} \triangleright \mathrm{id}_M"'] & & U \otimes (V \otimes (W \triangleright M)) \\
(U \otimes (V \otimes W)) \triangleright M \arrow[rr, "a^{\triangleright}_{U,V\otimes W,M}"] & & U \otimes ((V \otimes W) \triangleright M) \arrow[u, "\mathrm{id}_U \otimes a^{\triangleright}_{V,W,M}"']
\end{tikzcd}$$

and

$$\begin{tikzcd}
(V \otimes \mathbf{1}) \triangleright X \arrow[rr, "a^{\triangleright}_{V,\mathbf{1},M}"] \arrow[dr, "r_V \triangleright \mathrm{id}_M"'] & & V \otimes (\mathbf{1} \triangleright M) \arrow[dl, "\mathrm{id}_V \triangleright \lambda_M"] \\
& V \triangleright M &
\end{tikzcd}$$

for $U, V, W \in \mathrm{Obj}(\mathcal{C})$, and $M \in \mathrm{Obj}(\mathcal{M})$.

The natural isomorphism $a^{\triangleright}$ is called the *module constraint*, or also the mixed associator. The pentagon and triangle axioms are straightforward modifications of those of a monoidal category (see Definition 2.13). Right module categories over a monoidal category and bimodule categories over a pair of monoidal categories are defined analogously. For the latter, in addition to the module constraints for the left and right actions individually, there are two further pentagon identities involving also a natural family of isomorphisms $(X \triangleright M) \triangleleft Y \xrightarrow{\cong} X \triangleright (M \triangleleft Y)$; see, e.g., [**EtGNO**, Def. 7.1.7].

The *direct sum* $\mathcal{A} \oplus \mathcal{B}$ of two additive categories $\mathcal{A}$ and $\mathcal{B}$ is defined as follows. An object in $\mathcal{A} \oplus \mathcal{B}$ is a pair (a, b) consisting of an object $a \in \mathcal{A}$ and an object $b \in \mathcal{B}$, and the abelian morphism group $\mathrm{Hom}_{\mathcal{A}\oplus\mathcal{B}}((a,b),(a',b'))$ is given by $\mathrm{Hom}_{\mathcal{A}}(a,a') \oplus \mathrm{Hom}_{\mathcal{B}}(b,b')$. (In the latter expression the symbol $\oplus$ denotes the direct sum of abelian groups when the additive categories $\mathcal{A}$ and $\mathcal{B}$ are not enriched, while it stands for the direct sum of $\mathbb{K}$-vector spaces when they are enriched over $\mathcal{V}ect(\mathbb{K})$.)

Given two additive module categories $\mathcal{M}_1$ and $\mathcal{M}_2$ over a monoidal category $\mathcal{C}$, endowing their direct sum $\mathcal{M}_1 \oplus \mathcal{M}_2$ with a $\mathcal{C}$-action and with module constraints being sums of those of $\mathcal{M}_1$ and $\mathcal{M}_2$ yields again a $\mathcal{C}$-module category. An *indecomposable* module category is one that is not equivalent as a module category to such a direct sum $\mathcal{M}_1 \oplus \mathcal{M}_2$ with nonzero $\mathcal{M}_1$ and $\mathcal{M}_2$.

EXAMPLE 2.37. It is, in general, algorithmically hard to classify module categories. For instance, the indecomposable module categories over the monoidal category $\mathcal{V}ect_G(\mathbb{C})$ of G-graded vector complex spaces (as introduced in Example 2.15(5)) are classified by conjugacy classes of pairs (H, ω) consisting of a subgroup $H \leq G$ and a class $\omega \in H^2(H, \mathbb{C}^\times)$ in group cohomology. This generalizes to categories $\mathcal{V}ect_G^\omega(\mathbb{C})$ in which the associator is twisted by a 3-cocycle of G (as described in Example 2.47 below); see, e.g., Example 2.1 in [**Ost**]. There is also a classification of indecomposable module categories over the Drinfeld center (to be defined in Remark 5.47) of $\mathcal{V}ect_G^\omega(\mathbb{C})$; see Theorem 3.6 of [**Ost**].

EXAMPLE 2.38. In some special cases a classification of module categories over a given monoidal category can be achieved. One such class of monoidal categories consists of finitely semisimple monoidal categories (fusion categories, to be precise, a notion to be introduced in Definition 5.49) that are based on the simple Lie algebra $\mathfrak{sl}(2)$. Their module categories can be classified because their tensor product has a particularly simple structure (and tensors describing it have special eigenvalues). The result exhibits an A-D-E type pattern [**KirO**].

REMARK 2.39. We refrain from a detailed description of applications of module categories, because this typically requires additional knowledge. Instead we give a few (biased) pointers to the literature: Pivotal module categories over spherical fusion categories describe boundary conditions in three-dimensional topological field theories of Turaev-Viro type [**FSV1**]. A modular tensor category $\mathcal{C}$ captures crucial aspects of a chiral conformal field theory; a pivotal module category over $\mathcal{C}$ describes a full, local conformal field theory; see [**FSWY**] for a review. Fusion

2-categories, which are ingredients in the construction of four-dimensional topological field theories, are built from a fusion category and module categories over it, too [**DoR**]. Finally, in so-called tensor network models, module categories over spherical fusion categories are the source of particular symmetries; see Section 6.4.2 for a few more details.

2.4. Monoidal functors and natural transformations

For monoidal categories, functors and natural transformations must be endowed with further structure and properties.

DEFINITION 2.40. Let $(\mathcal{C}, \otimes_\mathcal{C}, \mathbf{1}_\mathcal{C}, a_\mathcal{C}, l_\mathcal{C}, r_\mathcal{C})$ and $(\mathcal{D}, \otimes_\mathcal{D}, \mathbf{1}_\mathcal{D}, a_\mathcal{D}, l_\mathcal{D}, r_\mathcal{D})$ be monoidal categories. (In some of the formulas below we omit the subscripts that indicate the category to which the data belong.)

(1) A *monoidal functor*, or *tensor functor*, from $\mathcal{C}$ to $\mathcal{D}$ is a triple $(F, \varphi_0, \varphi_2)$ consisting of the following data:

- a functor $F\colon \mathcal{C} \to \mathcal{D}$;
- an isomorphism
$$\varphi_0 : \quad \mathbf{1}_\mathcal{D} \longrightarrow F(\mathbf{1}_\mathcal{C})$$
in the category $\mathcal{D}$;
- a natural isomorphism
$$\varphi_2 : \quad \otimes_\mathcal{D} \circ (F \times F) \longrightarrow F \circ \otimes_\mathcal{C}$$
of functors from $\mathcal{C} \times \mathcal{C}$ to $\mathcal{D}$.

  Thus in particular there is an isomorphism
$$\varphi_2(U,V) : \quad F(U) \otimes_\mathcal{D} F(V) \xrightarrow{\cong} F(U \otimes_\mathcal{C} V)$$
for every pair of objects $U, V \in \mathcal{C}$.

These data are required to satisfy constraints that can be expressed as the commutativity of the following diagrams:

- Compatibility with the associativity constraint:

$$\begin{array}{ccc}
\big(F(U) \otimes F(V)\big) \otimes F(W) & \xrightarrow{a_{F(U),F(V),F(W)}} & F(U) \otimes \big(F(V) \otimes F(W)\big) \\
\Big\downarrow {\scriptstyle \varphi_2(U,V) \otimes \mathrm{id}_{F(W)}} & & \Big\downarrow {\scriptstyle \mathrm{id}_{F(U)} \otimes \varphi_2(V,W)} \\
F(U \otimes V) \otimes F(W) & & F(U) \otimes F(V \otimes W) \\
\Big\downarrow {\scriptstyle \varphi_2(U \otimes V, W)} & & \Big\downarrow {\scriptstyle \varphi_2(U, V \otimes W)} \\
F((U \otimes V) \otimes W) & \xrightarrow{F(a_{U,V,W})} & F(U \otimes (V \otimes W))
\end{array}$$

- Compatibility with the left unit constraint:

$$\begin{array}{ccc}
\mathbf{1}_\mathcal{D} \otimes F(U) & \xrightarrow{l_{F(U)}} & F(U) \\
\Big\downarrow {\scriptstyle \varphi_0 \otimes \mathrm{id}_{F(U)}} & & \Big\uparrow {\scriptstyle F(l_U)} \\
F(\mathbf{1}_\mathcal{C}) \otimes F(U) & \xrightarrow{\varphi_2(\mathbf{1}_\mathcal{C}, U)} & F(\mathbf{1}_\mathcal{C} \otimes U)
\end{array}$$

- Compatibility with the right unit constraint:

$$\begin{array}{ccc} F(U)\otimes \mathbf{1}_{\mathcal{D}} & \xrightarrow{\;r_{F(U)}\;} & F(U) \\ {\scriptstyle \mathrm{id}_{F(U)}\otimes\varphi_0}\downarrow & & \uparrow{\scriptstyle F(r_U)} \\ F(U)\otimes F(\mathbf{1}_{\mathcal{C}}) & \xrightarrow{\;\varphi_2(U,\mathbf{1}_{\mathcal{C}})\;} & F(U\otimes \mathbf{1}_{\mathcal{C}}) \end{array}$$

(2) A monoidal functor is called *strict* if the isomorphism φ_0 and the natural transformation φ_2 are identities in $\mathcal{D}$.

REMARK 2.41. A monoidal functor as given by Definition 2.40 is sometimes instead called a *strong monoidal functor*. There is a variant of the notion for which φ_0 and $\varphi_2(U,V)$ are not required to be isomorphisms, but are allowed to be non-invertible, with the rest of the definition unchanged. In this case the functor is called *lax monoidal*. Analogously, for an *oplax monoidal* functor one has morphisms in the direction opposite to the one of φ_0 and $\varphi_2(U,V)$ that are not required to be isomorphisms. In this text, monoidal functors are to be understood in the sense of Definition 2.40, i.e., as being strong, unless stated otherwise.

The composite $G\circ F$ of the functors $F\colon \mathcal{C}\longrightarrow\mathcal{D}$ and $G\colon \mathcal{D}\longrightarrow\mathcal{E}$ that underlie two monoidal functors $(F,\varphi_0^F,\varphi_2^F)$ and $(G,\varphi_0^G,\varphi_2^G)$ carries a natural monoidal structure, given by

$$\varphi_0^{G\circ F} = G(\varphi_0^F)\circ\varphi_0^G \qquad\text{and}\qquad \varphi_2^{G\circ F} = \varphi_2^G\circ G(\varphi_2^F)\,. \tag{2.42}$$

REMARK 2.43. Adjunctions are not automatically compatible with monoidal structures, but a lot of useful information about their compatibility is available. Pertinent statements can be derived systematically using the *doctrinal adjunctions* developed in [**Kel**].

EXERCISE 2.44. Let $\mathcal{C}$ and $\mathcal{D}$ be monoidal categories and $F\colon \mathcal{C}\longrightarrow\mathcal{D}$ be a monoidal functor. (Recall from Remark 2.41 that according to our conventions the latter means that the transformations φ_0 and φ_2 are isomorphisms.)

Show that a left adjoint $G\colon \mathcal{D}\longrightarrow\mathcal{C}$ of F, if it exists, comes with an oplax monoidal structure, while a right adjoint of F, if it exists, comes with a lax monoidal structure.

Hint: Use the description of adjunctions in terms of unit and counit.

DEFINITION 2.45.

(1) A *monoidal natural transformation*

$$\eta:\quad (F,\varphi_0,\varphi_2)\longrightarrow(F',\varphi_0',\varphi_2')$$

between monoidal functors F and F' from $\mathcal{C}$ to $\mathcal{D}$ is a natural transformation $\eta\colon F\longrightarrow F'$ with the following two compatibility properties:

- The diagram

$$\begin{array}{ccc} & & F(\mathbf{1}_{\mathcal{C}}) \\ & \nearrow^{\varphi_0} & \\ \mathbf{1}_{\mathcal{D}} & & \downarrow{\scriptstyle \eta_{\mathbf{1}}} \\ & \searrow_{\varphi_0'} & \\ & & F'(\mathbf{1}_{\mathcal{C}}) \end{array}$$

in $\mathcal{D}$ involving the tensor units of $\mathcal{C}$ and $\mathcal{D}$ commutes.

- For every pair (U, V) of objects of $\mathcal{C}$ the diagram

$$\begin{array}{ccc} F(U) \otimes F(V) & \xrightarrow{\varphi_2(U,V)} & F(U \otimes V) \\ {\scriptstyle \eta_U \otimes \eta_V} \downarrow & & \downarrow {\scriptstyle \eta_{U \otimes V}} \\ F'(U) \otimes F'(V) & \xrightarrow{\varphi'_2(U,V)} & F'(U \otimes V) \end{array}$$

  in $\mathcal{D}$ commutes.

(2) A *monoidal natural isomorphism* is an invertible monoidal natural transformation.

(3) An equivalence of monoidal categories $\mathcal{C}$ and $\mathcal{D}$, or *monoidal equivalence*, is given by a pair of monoidal functors $F\colon \mathcal{C} \to \mathcal{D}$ and $G\colon \mathcal{D} \to \mathcal{C}$ together with monoidal natural isomorphisms

$$\eta: \quad \mathrm{id}_{\mathcal{D}} \longrightarrow F \circ G \qquad \text{and} \qquad \theta: \quad G \circ F \longrightarrow \mathrm{id}_{\mathcal{C}}\,.$$

The natural transformations that are obtained by the vertical and horizontal composition (as described in Definition 1.49) of monoidal natural transformations are again monoidal.

REMARK 2.46. Every monoidal category is monoidally equivalent to a strict monoidal category; for a proof see, e.g., [**Kas**, Ch. XI.5]. As already announced in Example 2.15(1), this result is often invoked so as to simplify notation by replacing a given monoidal category by an equivalent strict one. There exist in fact even stronger strictification results; see Theorem 1.2 of [**JS2**].

EXAMPLE 2.47. We can generalize the category $\mathcal{V}ect_G(\mathbb{K})$ of G-graded vector spaces described in Example 2.15(5) to monoidal categories $\mathcal{V}ect_G^{\omega}(\mathbb{K})$ as follows.

For G a finite group, a function

$$\omega: \quad G \times G \times G \longrightarrow \mathbb{K}^{\times}$$

that satisfies the closedness condition

$$\begin{aligned} \omega(g_2, g_3, g_4) \cdot \omega^{-1}(g_1 g_2, g_3, g_4) \cdot \omega(g_1, g_2 g_3, g_4) \\ \cdot \omega^{-1}(g_1, g_2, g_3 g_4) \cdot \omega(g_1, g_2, g_3) = 1 \end{aligned} \tag{2.48}$$

for every quadruple $g_1, g_2, g_3, g_4 \in G$ is called a 3-*cocycle* on G (or, more specifically, a 3-cocycle in the group cohomology of G with values in the abelian group $\mathbb{K}^{\times}$, seen as a trivial G-module).

For any such 3-cocycle ω we can equip $\mathcal{V}ect_G(\mathbb{K})$ with a different monoidal structure for which the associator is determined by the family

$$\begin{aligned} a_{g_1,g_2,g_3}: \quad (\mathbb{K}_{g_1} \otimes \mathbb{K}_{g_2}) \otimes \mathbb{K}_{g_3} &\longrightarrow \mathbb{K}_{g_1} \otimes (\mathbb{K}_{g_2} \otimes \mathbb{K}_{g_3}) \\ (u_{g_1} \otimes u_{g_2}) \otimes u_{g_3} &\longmapsto \omega(g_1, g_2, g_3) \cdot u_{g_1} \otimes (u_{g_2} \otimes u_{g_3}) \end{aligned}$$

of linear maps, where the symbol $\otimes$ means the tensor product of vector spaces. Here $\mathbb{K}_g$ is the G-graded vector space given by the ground field $\mathbb{K}$ in degree g and the zero vector space in every other degree. The pentagon axiom for the associator a then amounts to the closedness of the 3-cocycle ω. As we will see in Chapter 4, such a 3-cocycle provides a concrete example for a topological Lagrangian of the type mentioned in Example 2.11.

EXERCISE 2.49. Consider again the situation studied in Example 2.47. Show that cohomologous cocycles give rise to monoidally equivalent categories.

The notion of a monoidal functor categorifies the one of a homomorphism between monoids. When considering module categories, there is a corresponding notion that categorifies the one of a homomorphism between modules over a monoid.

DEFINITION 2.50. An *oplax module functor* from a left $\mathcal{C}$-module category $(\mathcal{M}, \triangleright, a^{\triangleright}, \lambda)$ to a left $\mathcal{C}$-module category $(\mathcal{M}', \triangleright', a^{\triangleright\prime}, \lambda')$ consists of a functor $F\colon \mathcal{M} \to \mathcal{M}'$ and a natural family of morphisms

$$\varphi_{X,M}: \quad F(X \triangleright M) \longrightarrow X \triangleright' F(M)$$

for $X \in \mathcal{C}$ and $M \in \mathcal{M}$, such that the diagrams

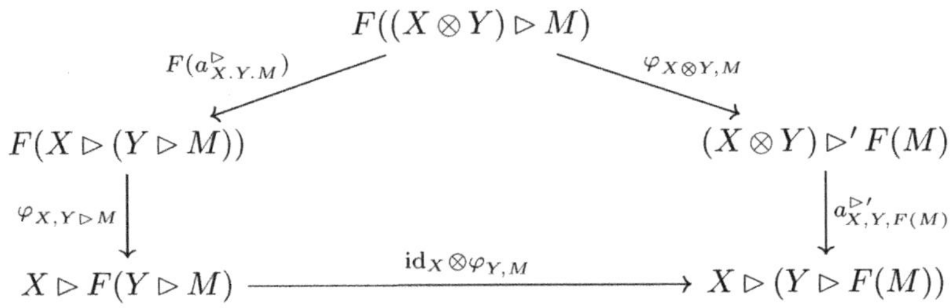

and

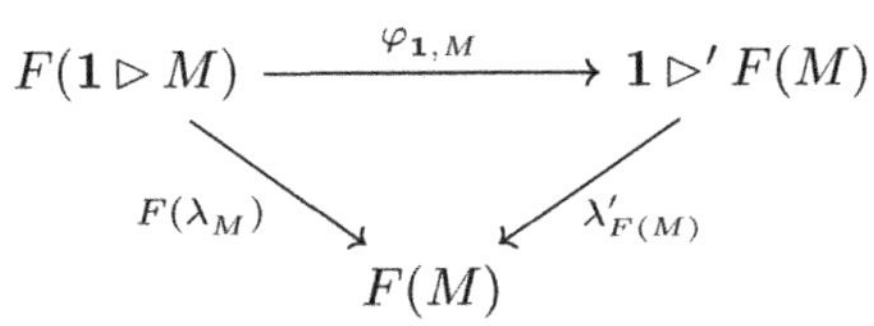

commute for all objects $X, Y \in \mathcal{C}$ and $M \in \mathcal{M}$.

If instead of the morphisms $\varphi_{X,M}$ there are morphisms in the opposite direction, satisfying analogous coherence conditions, then one deals instead with a *lax module functor*. And if all morphisms $\varphi_{X,M}$ are isomorphisms, then F is called a *strong module functor*, or just module functor.

2.5. Rigid monoidal categories

To proceed, we introduce a further property that objects in a monoidal category can have. This imitates a salient feature of finite-dimensional vector spaces. Specifically, the subsequent discussion of vector spaces motivates Definition 2.53 below.

For any $\mathbb{K}$-vector space X there are two pairs of maps that relate X and its linear dual space

$$X^* = \mathrm{Hom}_{\mathbb{K}}(X, \mathbb{K}) .$$

(1) Without any condition on the vector space X, we have two *evaluation maps*

$$\begin{aligned} d_X : \quad X^* \otimes X &\longrightarrow \mathbb{K}, \\ \xi \otimes x &\longmapsto \xi(x) \end{aligned}$$

and

$$\begin{aligned} \widetilde{d}_X : \quad X \otimes X^* &\longrightarrow \mathbb{K}, \\ x \otimes \xi &\longmapsto \xi(x) . \end{aligned}$$

We call d_X a *right evaluation* and $\widetilde{d}_X$ a *left evaluation*.

(2) Provided that the vector space X is *finite-dimensional*, we can further define two maps in the opposite direction, accordingly referred to as *co*evaluation maps. This can be done with the help of a basis $\{x_i\}_{i\in I}$ of X and the associated dual basis $\{\xi^i\}_{i\in I}$ of X^* (which, by definition, satisfies $\xi^i(x_j)=\delta^i_j$):

$$b_X: \quad \mathbb{K} \longrightarrow X \otimes X^*, \qquad \lambda \longmapsto \lambda \textstyle\sum_{i\in I} x_i \otimes \xi^i$$

and

$$\widetilde{b}_X: \quad \mathbb{K} \longrightarrow X^* \otimes X, \qquad \lambda \longmapsto \lambda \textstyle\sum_{i\in I} \xi^i \otimes x_i .$$

The maps b_X and $\widetilde{b}_X$ are actually independent of the choice of (dual) bases. This is seen by writing b_X in the form

$$b_X: \quad \mathbb{K} \longrightarrow \mathrm{End}_{\mathbb{K}}(X) \cong X \otimes X^*, \qquad \lambda \longmapsto \lambda\, \mathrm{id}_X ,$$

and analogously for $\widetilde{b}_X$.

We call b_X a *right coevaluation* and $\widetilde{b}_X$ a *left coevaluation*.

Suppressing the associators on the monoidal category of vector spaces, we compute

$$\begin{aligned}(\mathrm{id}_X \otimes d_X) \circ (b_X \otimes \mathrm{id}_X)(x) &= (\mathrm{id}_X \otimes d_X)\Big(\sum_{i\in I} x_i \otimes \xi^i \otimes x\Big) \\ &= \sum_{i\in I} \big(\xi^i(x)\big)\, x_i = x\end{aligned}$$

for any finite-dimensional vector space X. We thus conclude that

$$(\mathrm{id}_X \otimes d_X) \circ (b_X \otimes \mathrm{id}_X) = \mathrm{id}_X . \tag{2.51}$$

Similarly one shows that

$$(d_X \otimes \mathrm{id}_{X^*}) \circ (\mathrm{id}_{X^*} \otimes b_X) = \mathrm{id}_{X^*}. \tag{2.52}$$

One may wonder whether an analogous structure also exists for objects in a category of cobordisms. This is indeed the case. Recall that together with any oriented manifold M we also have another oriented manifold $\overline{M}$, which is the same manifold but with reversed orientation. Accordingly, for a $(d-1)$-dimensional manifold M, the d-manifold $M \times [-1,1]$, i.e., the cylinder over M, can be interpreted as a cobordism in six different ways, depending on whether its boundary components are regarded as incoming or outgoing and whether the embedding of M into the cylinder that maps it to a boundary component is an orientation preserving or orientation reversing map:

- as the identity cobordism
$$\mathrm{id}_M: \quad M \longrightarrow M;$$
- as an evaluation cobordism
$$d_M: \quad \overline{M} \sqcup M \longrightarrow \emptyset;$$
- as a coevaluation cobordism
$$b_M: \quad \emptyset \longrightarrow M \sqcup \overline{M};$$
- or as the cobordism obtained from any of these when replacing M by $\overline{M}$.

In the particular case that M is a circle, one commonly displays these different possibilities in the following suggestive manner:

$\mathrm{id}_M =$ 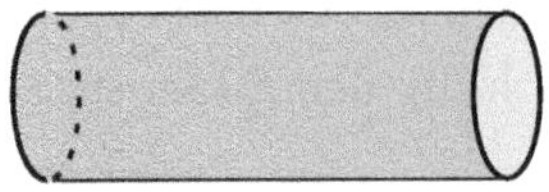

$d_M =$ 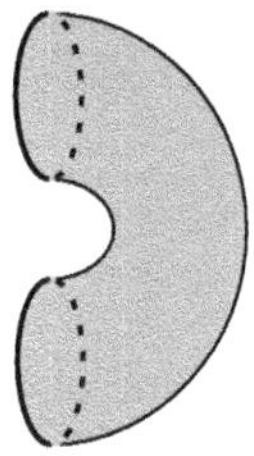 $b_M =$ 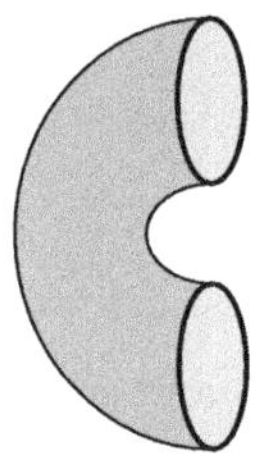

These morphisms satisfy the relations

$$(\mathrm{id}_M \otimes d_M) \circ (b_M \otimes \mathrm{id}_M) = \mathrm{id}_M$$

and

$$(d_M \otimes \mathrm{id}_M) \circ (\mathrm{id}_M \otimes b_M) = \mathrm{id}_M \,.$$

that are analogous to (2.51) and (2.52).

Graphically these identities look as follows:

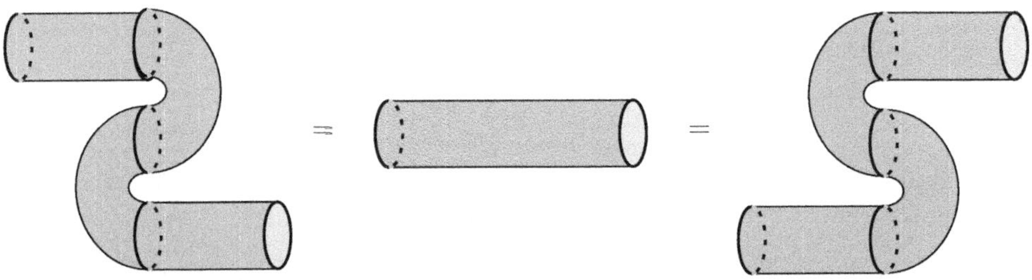

These relations – not only in the cobordism category – are sometimes referred to as *snake relations* because they amount to straightening a "snakelike" curved line. (Another common name for them is *zig-zag relations*.)

Asking for similar structures in general monoidal categories leads to the following notion:

DEFINITION 2.53.

(1) An object X of a monoidal category $\mathcal{C}$ is called *right dualizable* if there exists an object $Y \in \mathcal{C}$ and a pair

$$b_X : \quad \mathbf{1} \longrightarrow X \otimes Y \qquad \text{and} \qquad d_X : \quad Y \otimes X \longrightarrow \mathbf{1}$$

of morphisms that satisfy

$$(\mathrm{id}_X \otimes d_X) \circ a_{X,Y,X} \circ (b_X \otimes \mathrm{id}_X) = \mathrm{id}_X$$

and

$$(d_X \otimes \mathrm{id}_Y) \circ a^{-1}_{Y,X,Y} \circ (\mathrm{id}_Y \otimes b_X) = \mathrm{id}_Y \,.$$

Such an object is denoted by $Y =: X^\vee$ and is called a *right dual* to X.

The morphism d_X is called a (right) *evaluation*, and the morphism b_X a (right) *coevaluation*.

(2) A monoidal category is called *right-rigid*, or *right-autonomous*, if every object has a right dual.

(3) A *left dual* to $X \in \mathcal{C}$ is an object ${}^{\vee}X$ of $\mathcal{C}$ together with morphisms

$$\widetilde{b}_X : \quad \mathbf{1} \longrightarrow {}^{\vee}X \otimes X \qquad \text{and} \qquad \widetilde{d}_X : \quad X \otimes {}^{\vee}X \longrightarrow \mathbf{1}$$

such that

$$(\widetilde{d}_X \otimes \mathrm{id}_X) \circ a^{-1}_{X,{}^{\vee}X,X} \circ (\mathrm{id}_X \otimes \widetilde{b}_X) = \mathrm{id}_X$$

and

$$(\mathrm{id}_X \otimes \widetilde{d}_X) \circ a_{{}^{\vee}X,X,{}^{\vee}X} \circ (\widetilde{b}_X \otimes \mathrm{id}_X) = \mathrm{id}_{{}^{\vee}X} .$$

A *left-rigid*, or *left-autonomous*, category is a monoidal category in which every object has a left dual.

(4) A monoidal category is said to be *rigid*, or *autonomous*, if it is both left and right rigid.

REMARK 2.54. For clarity, in the formulas above we have explicitly included the associator a of the monoidal category $\mathcal{C} = (\mathcal{C}, \otimes, \mathbf{1}, a, l, r)$. Below, we will suppress a (as well as the unitors l and r) whenever their presence can easily be restored from the context.

Left and right duals are essentially unique, thus allowing us the speak of "the" left and right dual of an object:

LEMMA 2.55. Let V be an object in a monoidal category, and let $(V^{\vee}, d_V, b_V)$ and $(V^{\wedge}, d^{\wedge}_V, b^{\wedge}_V)$ be two right duals of V. Then $V^{\vee}$ and $V^{\wedge}$ are canonically isomorphic: There is a unique isomorphism $\varphi\colon V^{\vee} \to V^{\wedge}$ such that the two diagrams

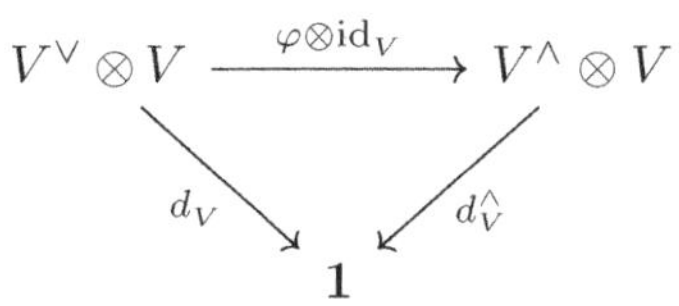

and

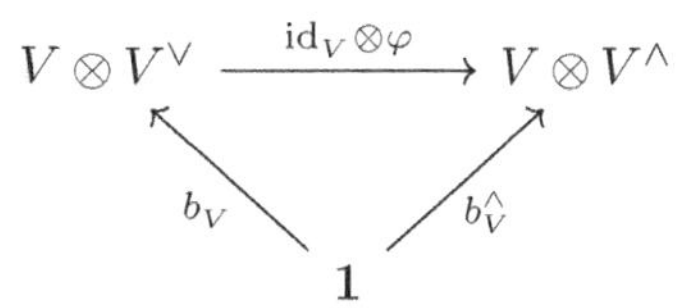

commute. An analogous result holds for left duals.

PROOF. For ease of notation, we assume that the monoidal category is strict. The axioms of a duality then directly imply that φ has to be given by

$$\varphi : \quad V^{\vee} \xrightarrow{\mathrm{id}_{V^{\vee}} \otimes b^{\wedge}_V} V^{\vee} \otimes V \otimes V^{\wedge} \xrightarrow{d_V \otimes \mathrm{id}_{V^{\wedge}}} V^{\wedge}.$$

This map is in fact an isomorphism: the morphism

$$\varphi^{-1} : \quad V^{\wedge} \xrightarrow{\mathrm{id}_{V^{\wedge}} \otimes b_V} V^{\wedge} \otimes V \otimes V^{\vee} \xrightarrow{d^{\wedge}_V \otimes \mathrm{id}_{V^{\vee}}} V^{\vee}$$

is a two-sided inverse.

Alternatively, the statement follows from the uniqueness of adjoints considered in Exercise 1.129. □

REMARK 2.56. Dualities can be taken to be compatible with the monoidal structure. More specifically, we can require that

$$\mathbf{1}^{\vee} = \mathbf{1} \qquad \text{and} \qquad (X \otimes Y)^{\vee} = Y^{\vee} \otimes X^{\vee}.$$

The first equality holds because we can take $b_{\mathbf{1}}$ and $d_{\mathbf{1}}$ to be given by the identity morphism. Concerning the second equality, note that the snake identities for the evaluations and coevaluations of X and Y directly imply that the morphisms

$$b_{X \otimes Y} = (\mathrm{id}_X \otimes b_Y \otimes \mathrm{id}_{X^{\vee}}) \circ b_X$$

and

$$d_{X \otimes Y} = d_Y \circ (\mathrm{id}_{Y^{\vee}} \otimes d_X \otimes \mathrm{id}_Y)$$

satisfy the snake identities as well, thus identifying $Y^\vee \otimes X^\vee$ as a possible right dual of $X \otimes Y$. The statement then follows by invoking the uniqueness result of Lemma 2.55. Also, with this choice we have

$$(a_{X,Y,Z})^\vee = a_{Z^\vee,Y^\vee,X^\vee} \,.$$

(Here $f^\vee$ denotes the right dual of a morphism f, see below.) Analogous formulas hold for a left duality.

EXAMPLE 2.57. The motivational considerations at the beginning of this section can now be rephrased as follows. The monoidal categories of finite-dimensional vector spaces and of cobordisms are both rigid.

EXERCISE 2.58. Prove that the monoidal category of finite-dimensional representations of a group G is rigid.

Hint: Given a finite-dimensional G-representation V, show that the linear dual V^* with action given by $(g\,.\,\alpha)(v) := \alpha(g^{-1}.\,v)$ for $v \in V$, $\alpha \in V^*$ and $g \in G$ is both a left and right dual of the G-representation V.

In a right-rigid monoidal category $\mathcal{C}$ the assignment of the dual object $X^\vee$ to every object X, together with the corresponding evaluation and coevaluation morphisms, is called a *right duality* for $\mathcal{C}$. We define an action of a right duality not only on objects, but also on morphisms, namely as

$$\begin{aligned} \mathrm{Hom}(X,Y) \ni f \\ \longmapsto f^\vee := (d_Y \otimes \mathrm{id}_{X^\vee}) \circ (\mathrm{id}_{Y^\vee} \otimes f \otimes \mathrm{id}_{X^\vee}) \circ (\mathrm{id}_{Y^\vee} \otimes b_X) \\ \in \mathrm{Hom}(Y^\vee, X^\vee)\,. \end{aligned}$$

This prescription yields a functor

$$(-)^\vee : \quad \mathcal{C} \longrightarrow \mathcal{C}^{\mathrm{opp}},$$

i.e., a contravariant functor. Similarly, for a left-rigid monoidal category $\mathcal{C}$, we get a left duality functor

$${}^\vee(-) : \quad \mathcal{C} \longrightarrow \mathcal{C}^{\mathrm{opp}}.$$

EXERCISE 2.59. As seen in Example 2.15(2), the category $\mathcal{E}nd(\mathcal{C})$ of endofunctors of a small category $\mathcal{C}$ is a monoidal category.

Show that for an endofunctor $F\colon \mathcal{C} \to \mathcal{C}$ a left (right) dual in the monoidal category $\mathcal{E}nd(\mathcal{C})$ is precisely a left (right) adjoint functor for F.

GRAPHICAL DESCRIPTION 2.60. In the graphical string calculus for monoidal categories, we represent the evaluation and coevaluation morphisms of left and right dualities by string diagrams as follows:

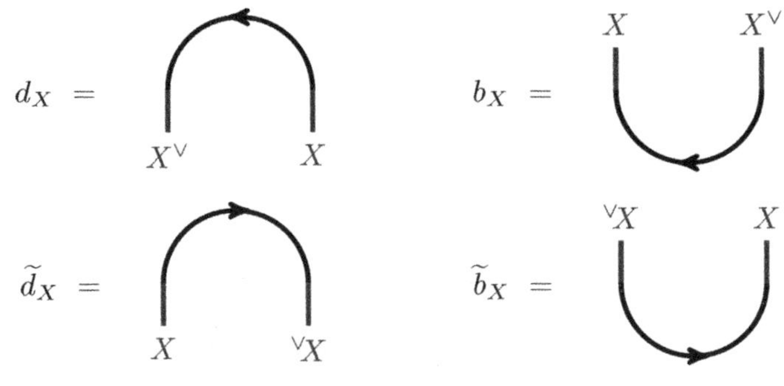

The defining snake relations for the right duality morphisms then look like

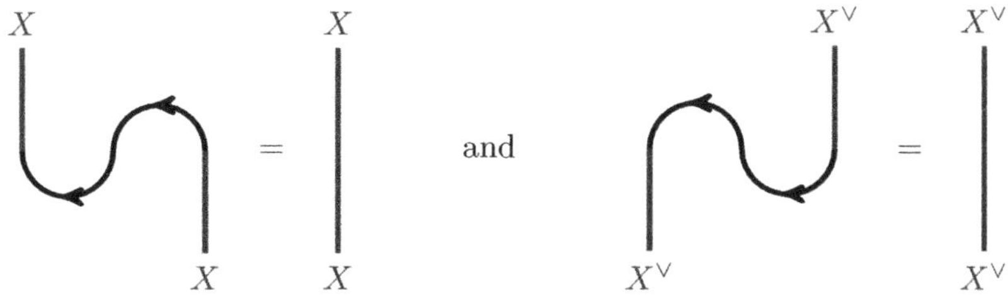

and similarly for the left duality morphisms.

The action on morphisms is graphically expressed as

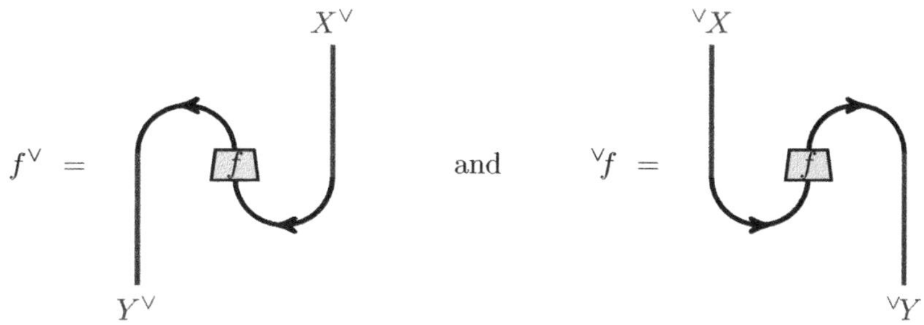

for $f \in \mathrm{Hom}(X, Y)$.

REMARK 2.61. A closed graph embedded in $\mathbb{R}^2$ whose edges and vertices are labeled by objects and morphisms, respectively, in a rigid monoidal category $\mathcal{C}$ represents an endomorphism of the monoidal unit $\mathbf{1}$ of $\mathcal{C}$. In particular, if the object $\mathbf{1}$ is absolutely simple, i.e., $\mathrm{Hom}(\mathbf{1}, \mathbf{1}) \cong \mathbb{K}$, then the value of the graph can be interpreted as a number.

LEMMA 2.62. *A $\mathbb{K}$-vector space V admits a dual (that is, both* left *and* right *dual) if and only if it is finite-dimensional.*

PROOF. We present the proof for the right dual; the proof for the left dual is analogous. The element $b_V(1) \in V \otimes V^*$ can be written as a linear combination

$$b_V(1) = \sum_{i=1}^{N} b_i \otimes \beta^i$$

with suitable elements $b_i \in V$ and $\beta^i \in V^*$ (by definition, this is a finite sum). By the axioms of a duality it follows that

$$v = (\mathrm{id}_V \otimes d_V) \circ (b_V(1) \otimes \mathrm{id}_V)(v) = \sum_{i=1}^{N} b_i \, \beta^i(v)$$

for any vector $v \in V$. Thus the finitely many vectors $(b_i)_{i=1,\dots,N}$ form a generating system for V. This implies that V is finite-dimensional. The converse has already been established, through Equations (2.51) and (2.52), in the motivational considerations at the beginning of the present section. □

EXERCISE 2.63. Let $F\colon \mathcal{C} \to \mathcal{D}$ be a monoidal functor and let $X^\vee$ be a right dual of an object $X \in \mathcal{C}$. Show that by the evaluation and coevaluation morphisms of X and $X^\vee$ and the structure morphisms of F, the object $F(X^\vee)$ can be made into a right dual of $F(X)$. In other words, monoidal functors preserve right (and similarly, left) dualities.

REMARK 2.64. We will see that the functor $Z\colon \mathcal{C}ob_{d,d-1} \to \mathcal{V}ect(\mathbb{K})$ in our tentative definition of a topological field theory in Example 1.58 should be equipped with the structure of a monoidal functor. Exercise 2.63, combined with with Lemma 2.62, then implies that for every closed oriented $(d-1)$-manifold M the vector space $Z(M)$ is finite-dimensional. Its dual $Z(M)^\vee$ is canonically isomorphic to $Z(\overline{M})$, where $\overline{M}$ is the closed oriented $(d-1)$-manifold that is obtained from M by orientation reversal.

EXERCISE 2.65. Show that for any rigid category $\mathcal{C}$ the *double (right) dual functor* $(-)^{\vee\vee}\colon \mathcal{C} \to \mathcal{C}$ comes with the structure of a monoidal functor. (The same is true for the double left dual functor ${}^{\vee\vee}(-)$.)

Hint: Recall from Remark 2.56 that $d_{X\otimes Y} = d_Y \circ (\mathrm{id}_{Y^\vee} \otimes d_X \otimes \mathrm{id}_Y)$.

There is no reason, in general, for the right and left duality functors of a rigid monoidal category to be isomorphic. This does, however, hold for *pivotal* categories. In order to introduce this notion, we invoke the monoidal structure on the double dual functor $(-)^{\vee\vee}\colon \mathcal{C} \to \mathcal{C}$ from Exercise 2.65.

DEFINITION 2.66. A *pivotal structure* on a rigid monoidal category $\mathcal{C}$ is a monoidal natural isomorphism

$$\pi: \quad (-)^{\vee\vee} \xrightarrow{\cong} \mathrm{id}_{\mathcal{C}}$$

between the double dual and the identity functor.

A *pivotal category* is a rigid monoidal category together with a choice of a pivotal structure.

A rigid monoidal category need not admit a pivotal structure, and if it does, the pivotal structure need not be unique. Thus pivotality indeed amounts to structure and is not just a property.

EXAMPLE 2.67. Recall from Exercise 1.63 that for any finite-dimensional vector space V there is a canonical isomorphism $V^{**} \xrightarrow{\cong} V$. These isomorphisms endow the category of finite-dimensional vector spaces with a pivotal structure.

REMARK 2.68. The strictification result for monoidal categories stated in Remark 2.46 extends to the pivotal case: Every pivotal monoidal category is equivalent, as a pivotal monoidal category, to a strict pivotal monoidal category, i.e., one for which the pivotal structure is the identity natural transformation (Theorem 2.2 in [**NgS1**]).

EXERCISE 2.69. Let $\mathcal{C}$ be a pivotal category and X an object in $\mathcal{C}$. Show that the pivotal structure provides a canonical isomorphism between $X^\vee$ and ${}^\vee X$ and, more generally, a natural isomorphism $(-)^\vee \xrightarrow{\cong} {}^\vee(-)$ of functors. In other words, in a pivotal category left and right duality coincide up to canonical isomorphism.

The presence of a pivotal structure allows one to extend the notion of the trace of a linear map to the setting of monoidal categories:

DEFINITION 2.70. Let $f \in \mathrm{Hom}(X,X)$ be an endomorphism in a pivotal category $\mathcal{C}$ with pivotal structure π. The morphisms

$$\mathrm{tr_r}(f) := d_X \circ \big(\mathrm{id}_{X^\vee} \otimes (f \circ \pi_X^{-1})\big) \circ b_{X^\vee} \in \mathrm{Hom}(\mathbf{1},\mathbf{1})$$

and

$$\mathrm{tr_l}(f) := d_{X^\vee} \circ \big((\pi_X \circ f) \otimes \mathrm{id}_{X^\vee}\big) \circ b_X \in \mathrm{Hom}(\mathbf{1},\mathbf{1})$$

are called the *right trace* and *left trace* of f, respectively.

The *right* and *left dimension* of an object $X \in \mathcal{C}$ are the morphisms

$$\dim_\mathrm{r}(X) := \mathrm{tr_r}(\mathrm{id}_X) \qquad \text{and} \qquad \dim_\mathrm{l}(X) := \mathrm{tr_l}(\mathrm{id}_X)\,,$$

respectively.

Note that the traces, and thus in particular the dimensions, do depend on the choice of pivotal structure, even though this is not indicated in the notation. If $\mathcal{C}$ is in addition $\mathbb{K}$-linear and $\mathrm{Hom}(\mathbf{1},\mathbf{1}) \cong \mathbb{K}$, it is common to invoke this isomorphism to interpret traces as elements of the field $\mathbb{K}$.

EXERCISE 2.71.

(1) Show that the traces $\mathrm{tr_r}$ and $\mathrm{tr_l}$ are *cyclic* in the sense that $\mathrm{tr_{r,l}}(g \circ f) = \mathrm{tr_{r,l}}(f \circ g)$ for any pair of morphisms $f \in \mathrm{Hom}(X,Y)$ and $g \in \mathrm{Hom}(Y,X)$.
(2) Show that $\mathrm{tr_{r,l}}(f^\vee) = \mathrm{tr_{l,r}}(f)$ for any endomorphism f.
(3) Assume that the monoidal unit is absolutely simple, i.e., obeys $\mathrm{Hom}(\mathbf{1},\mathbf{1}) = \mathbb{K}\,\mathrm{id}_\mathbf{1}$. Conclude that the traces $\mathrm{tr_r}$ and $\mathrm{tr_l}$ are multiplicative in the sense that
$$\mathrm{tr_{r,l}}(f \otimes f') = \mathrm{tr_{r,l}}(f)\,\mathrm{tr_{r,l}}(f')$$
for any pair of morphisms f and f'.

REMARK 2.72. In the absence of a pivotal structure one can still define an analogous right trace for morphisms in $\mathrm{Hom}(X^{\vee\vee},X)$ and a left trace for morphisms in $\mathrm{Hom}(X,X^{\vee\vee})$.

DEFINITION 2.73. A *spherical structure* on a rigid monoidal category $\mathcal{C}$ is a pivotal structure for which the left and right traces of any endomorphism are equal.

When describing morphisms in a spherical category in terms of string diagrams, the canvas can be taken to be a two-sphere instead of a rectangle in the plane; this is the origin of the terminology 'spherical'.

2.6. Hopf algebras

Recall from Theorem 2.33 that the category of modules over a bialgebra (as defined in Definition 2.28) has the structure of a monoidal category. Since finite-dimensional vector spaces are rigid, it is reasonable to ask for what bialgebras their finite-dimensional modules are still rigid. The answer to this question involves the following concepts:

DEFINITION 2.74. Let $A = (A,\mu,\eta,\Delta,\varepsilon)$ be a bialgebra over a field $\mathbb{K}$.

(1) An *antipode* on A is a $\mathbb{K}$-linear map $\mathrm{s}\colon A \to A$ satisfying
$$\mu \circ (\mathrm{s} \otimes \mathrm{id}_A) \circ \Delta = \eta \circ \varepsilon = \mu \circ (\mathrm{id}_A \otimes \mathrm{s}) \circ \Delta\,.$$

(2) A *skew antipode* on A is a $\mathbb{K}$-linear map $\widetilde{\mathrm{S}}\colon A \to A$ satisfying

$$\mu \circ \tau_{A,A} \circ (\widetilde{\mathrm{S}} \otimes \mathrm{id}_A) \circ \Delta = \eta \circ \varepsilon = \mu \circ \tau_{A,A} \circ (\mathrm{id}_A \otimes \widetilde{\mathrm{S}}) \circ \Delta ,$$

with $\tau_{A,A}$ the flip map (2.21).

DEFINITION 2.75. For $\mathbb{K}$ a field, a *Hopf algebra* $H = (H, \mu, \eta, \Delta, \varepsilon, \mathrm{S})$ over $\mathbb{K}$ is a $\mathbb{K}$-bialgebra $(H, \mu, \eta, \Delta, \varepsilon)$ endowed with an antipode $\mathrm{S}\colon H \longrightarrow H$.

EXAMPLE 2.76. For G a group and $\mathbb{K}$ a field, consider the group algebra $\mathbb{K}[G]$ with basis $\{\beta_g \mid g \in G\}$, as introduced in Definition 1.15. Setting

$$\varepsilon(\beta_g) := \delta_{g,e} , \qquad \Delta(\beta_g) := \beta_g \otimes \beta_g \qquad \text{and} \qquad \mathrm{S}(\beta_g) := \beta_{g^{-1}}$$

for all $g \in G$ (with g^{-1} the inverse of g in G) endows $\mathbb{K}[G]$ with the structure of a Hopf algebra.

EXERCISE 2.77. Show:

(1) If a bialgebra admits an antipode S, then S is unique.

(2) If an antipode S is invertible, then its inverse S^{-1} is a skew antipode.

(3) The sixtuple $(H, \mu, \eta, \Delta, \varepsilon, \widetilde{\mathrm{S}})$ is a bialgebra with a skew antipode if and only if

$$(H, \mu \circ \tau_{H,H}, \eta, \Delta, \varepsilon, \widetilde{\mathrm{S}})$$

is a Hopf algebra, and if and only if

$$(H, \mu, \eta, \tau_{H,H} \circ \Delta, \varepsilon, \widetilde{\mathrm{S}})$$

is a Hopf algebra.

REMARK 2.78. The antipode of any commutative or cocommutative Hopf algebras squares to the identity. The antipode of any finite-dimensional Hopf algebra is invertible (see, e.g., Corollary 5.1.6 in [**Swe**]). Hopf algebras with non-invertible antipode are in fact not very easy to construct; one possible construction is described in [**Tak**]. Accordingly, sometimes, e.g., in Definition 5.3.10 of [**EtGNO**], invertibility of the antipode is included in the notion of a Hopf algebra.

EXERCISE 2.79. Consider the commutative algebra $\mathcal{A}_n(\mathbb{K}) := \mathbb{K}[X_{i,j} \mid 1 \leq i, j \leq n]$ of polynomials in n^2 indeterminates $\{X_{i,j}\}_{1 \leq i,j \leq n}$.
By setting

$$\Delta(X_{i,j}) := \sum_{k=1}^{n} X_{i,k} \otimes X_{k,j} \qquad \text{and} \qquad \varepsilon(X_{i,j}) := \delta_{i,j}$$

we obtain two algebra morphisms

$$\Delta : \quad \mathcal{A}_n(\mathbb{K}) \longrightarrow \mathcal{A}_n(\mathbb{K}) \otimes \mathcal{A}_n(\mathbb{K}) \qquad \text{and} \qquad \varepsilon : \quad \mathcal{A}_n(\mathbb{K}) \longrightarrow \mathbb{K}$$

(1) Show that this defines a bialgebra structure on $\mathcal{A}_n(\mathbb{K})$.

(2) Denote by $X := \big(X_{i,j}\big)_{1 \leq i,j \leq n}$ the $n \times n$-matrix whose entries are given by the indeterminates $X_{i,j} \in \mathcal{A}_n(\mathbb{K})$.

Notice that the Leibniz formula $\det(M) = \sum_{\sigma \in \mathfrak{S}_n} \mathrm{sign}(\sigma) \prod_{j=1}^{n} M_{j,\sigma(j)}$ for the determinant of an $n \times n$-matrix, in which M is usually taken to have entries $M_{i,j}$ that are elements of a field, still makes sense for matrices with entries in an arbitrary commutative ring. Accordingly, in the present situation we can set $g := \det(X) \in \mathcal{A}_n(\mathbb{K})$.

Show that the element $g \in \mathcal{A}_n(\mathbb{K})$ is *group-like*, i.e., $\Delta(g) = g \otimes g$.

(3) Show that the bialgebra $\mathcal{A}_n(\mathbb{K})$ is not a Hopf algebra.

(4) Consider the two-sided ideal $\mathcal{I}$ of $\mathcal{A}_2(\mathbb{K})$ that is generated by $\det(X)-1$. Show that the quotient $\mathcal{A}_2(\mathbb{K})/\mathcal{I}$ is a Hopf algebra, with antipode given by $\mathrm{S}(X_{i,j}+\mathcal{I})=(X^{\#})_{i,j}+\mathcal{I}$, with

$$X^{\#} := \begin{pmatrix} X_{2,2} & -X_{1,2} \\ -X_{2,1} & X_{1,1} \end{pmatrix}.$$

How can one generalize this construction to $\mathcal{A}_n(\mathbb{K})$ for $n\geq 3$?

EXERCISE 2.80. One definition (out of various different ones that are in use) of a *quantum group* is that it is a non-commutative non-cocommutative Hopf algebra (rather than, as the term may suggest, a group) with additional structure. Important classes of quantum groups can be obtained by certain types of deformations of the universal enveloping algebra $\mathrm{U}(\mathfrak{g})$ (which we described in Example 2.24) of a Lie algebra $\mathfrak{g}$. The following example of a Hopf algebra arises as the Borel subalgebra of one variant of the quantum group $\mathrm{U}_q(\mathfrak{sl}(2))$.

Let $\mathbb{K}$ be a field and $q\in\mathbb{K}^{\times}$. Let H be the free algebra on the three generators x, x^{-1} and y modulo the relations

$$xy = q\,yx \qquad \text{and} \qquad xx^{-1} = 1 = x^{-1}x\,.$$

(1) Show that setting

$$\Delta(x^{\pm1}) := x^{\pm1}\otimes x^{\pm1}\,, \qquad \Delta(y) := y\otimes 1 + x\otimes y$$
$$\text{and} \qquad \varepsilon((x^{\pm1}) := 0\,, \qquad \varepsilon(y) := 0$$

yields algebra homomorphisms $\Delta\colon H\to H\otimes H$ and $\varepsilon\colon H\to\mathbb{K}$.

(2) Show that this turns H into a bialgebra, and that this bialgebra is a Hopf algebra.

(3) Compute the order of the antipode of H.

EXERCISE 2.81. Recall the Taft bialgebra H_N from Exercise 2.32.
Show that H_N admits an antipode S determined by

$$\mathrm{S}(g)=g^{-1} \quad\text{and}\quad \mathrm{S}(x)=-xg^{-1}.$$

Compute the order of S.

GRAPHICAL DESCRIPTION 2.82. Recall from Graphical Description 2.30 the string diagrams for the (co)unit and (co)multiplication of a bialgebra. Representing in addition the antipode graphically as

$\mathrm{S} =:$ 

the antipode axioms read

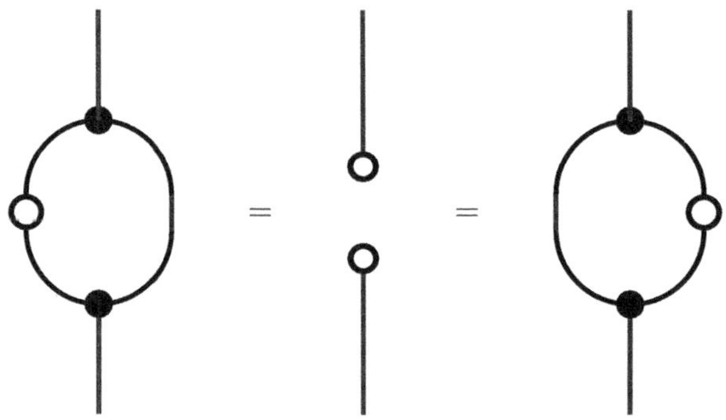

EXERCISE 2.83. For any coalgebra H over a field $\mathbb{K}$, the tensor product $H \otimes H$ inherits from H a natural structure of a coalgebra. As a consequence, if H is both a unital algebra and a counital coalgebra, then the $\mathbb{K}$-vector space of linear maps from $H \otimes H$ to H is a unital associative algebra with respect to the convolution product $\varphi \circledast \varphi' := \mu \circ (\varphi \otimes \varphi') \circ \Delta_{H \otimes H}$, with unit element $1_{\circledast} = \eta \circ (\varepsilon \otimes \varepsilon)$.

Assume that H is even a Hopf algebra. Show graphically that

$$(\mathrm{S} \circ \mu) \circledast \mu = 1_{\circledast} = \mu \circledast (\mu \circ \tau_{H,H} \circ (\mathrm{S} \otimes \mathrm{S})),$$

with $\tau_{H,H}$ the flip map (2.21). Conclude that $\mathrm{S} \circ \mu = \mu \circ \tau_{H,H} \circ (\mathrm{S} \otimes \mathrm{S})$, i.e., that the antipode of a Hopf algebra is an algebra antihomomorphism. Show further that the antipode is also an antihomomorphism of coalgebras.

We are now in a position to answer the question posed at the beginning of this subsection:

THEOREM 2.84.

(1) Let $(H, \mu, \eta, \Delta, \varepsilon, \mathrm{S})$ be a Hopf algebra. Then the category H-mod of finite-dimensional H-modules admits a right duality.

(2) Let $(H, \mu, \eta, \Delta, \varepsilon, \widetilde{\mathrm{S}})$ be a bialgebra with skew antipode. Then the category H-mod of finite-dimensional H-modules admits a left duality.

(3) Let $(H, \mu, \eta, \Delta, \varepsilon, \mathrm{S})$ be a Hopf algebra with invertible antipode. Then the category H-mod of finite-dimensional H-modules admits right and left dualities. Moreover, the right and left duals of each H-module are isomorphic if and only if the square S^2 is an inner automorphism of H.

One can in fact show that the sufficient conditions in these statements are also necessary. Concretely, a right dual of an H-module (X, ρ) is given by $(X^*, \rho_{X^\vee})$ with

$$\rho_{X^\vee} = (d_X \otimes \mathrm{id}_{X^*}) \circ (\mathrm{id}_{X^*} \otimes \rho \otimes \mathrm{id}_{X^*}) \circ (\mathrm{id}_{X^*} \otimes \mathrm{S} \otimes b_X) \circ \tau_{H,X^*},$$

while a left dual is given by $(X^*, \rho_{{}^\vee X})$ with

$$\begin{aligned} \rho_{{}^\vee X} = {} & (\mathrm{id}_{X^*} \otimes \widetilde{d}_X) \circ (\mathrm{id}_{X^*} \otimes \rho \otimes \mathrm{id}_{X^*}) \\ & \circ [(\mathrm{id}_{X^*} \otimes \mathrm{S}^{-1}) \circ \tau_{H,X^*}] \otimes \mathrm{id}_X \otimes \mathrm{id}_{X^*}) \circ (\mathrm{id}_H \otimes \widetilde{b} \otimes \mathrm{id}_{X^*}) \end{aligned}$$

with τ_{H,X^*} the flip map $h \otimes \varphi \mapsto \varphi \otimes h$ for $h \in H$ and $\varphi \in X^*$.

Graphical Description 2.85. Expressing the action of H on X by the string diagram

ρ_X =: 

the right dual action is given by

$\rho_{X^\vee}$ = 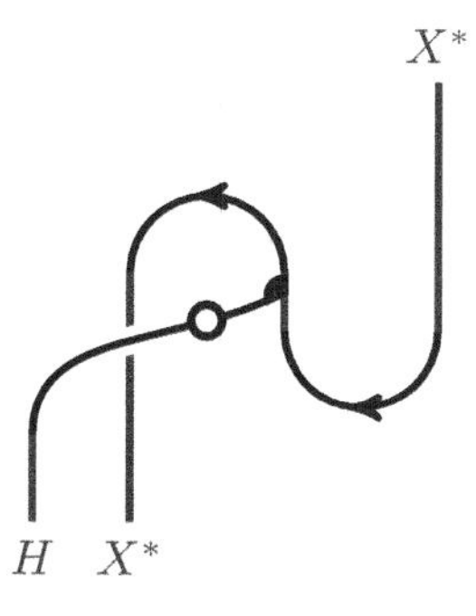

That this indeed endows the dual vector space X^* with the structure of an H-module is seen by noticing that

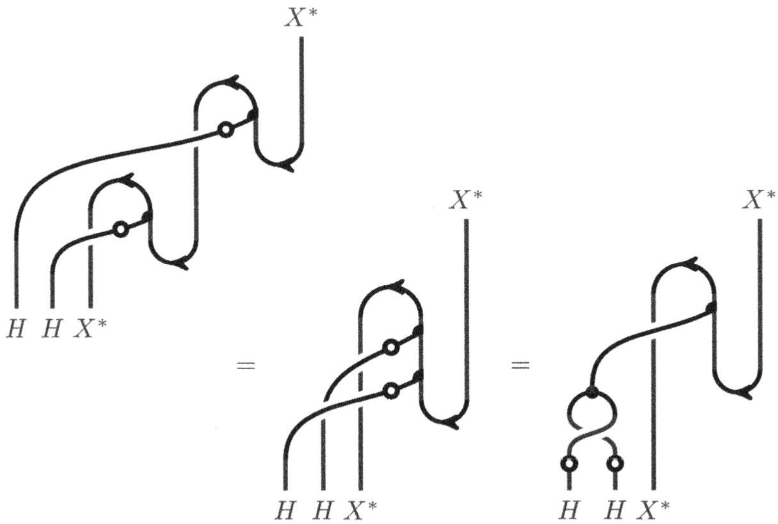

and then invoking the anti-algebra homomorphism property of the antipode.

REMARK 2.86. It is natural to ask whether the considerations in Exercise 2.83 and in Graphical Description 2.85 can be generalized to the situation that H is a Hopf algebra in a monoidal category $\mathcal{C}$. As already pointed out at the end of Graphical Description 2.30, the structure of a *braiding* c on $\mathcal{C}$, which provides a generalization of the flip map τ, makes it possible to define the notion of a bialgebra in $\mathcal{C}$; in fact it also allows one to define the notion of a Hopf algebra. Braidings will be introduced in the next section. If $\mathcal{C}$ is endowed with a braiding c, then one can, e.g., take over the definition of the right dual action $\rho_{X^\vee}$ by just replacing the flip map τ_{H,X^*} by the braiding morphism $c_{H,X^\vee}$, while one obtains a left dual action by replacing in $\rho_{^\vee X}$ the flip map by the opposite braiding $c^{-1}_{^\vee X,H}$.

The results in Theorem 2.84 motivate the notion of a *pivotal Hopf algebra* which, by definition, is a pair (H,ω) consisting of a Hopf algebra H and an element $\omega \in H$ that is group-like, i.e., obeys $\Delta(\omega) = \omega \otimes \omega$, and satisfies

$$S^2(x) = \omega\, x\, \omega^{-1}.$$

Such an element ω is called a *pivot* for H.

EXERCISE 2.87. Let H be a finite-dimensional Hopf algebra.

(1) Show that the set of group-like elements of H forms a group, and that this subset of H is linearly independent.

(2) Show that if ω is a pivot for H, then the action with ω endows the category of finite-dimensional H-modules with the structure of a pivotal category.

(3) Show that if the monoidal category of finite-dimensional H-modules admits a pivotal structure π, then the element $\pi_H(\mathbf{1}_H) \in H^{**} \cong H$ is a pivot for H (using the canonical isomorphism of a finite-dimensional vector space with its double dual to identify H^{**} with H).

2.7. Braided monoidal categories

With the notions of monoidal category and monoidal functor at our disposal, we are now in a position to formulate an improved version of our definition of a topological field theory.

Tentative Definition. A topological field theory is a *monoidal* functor

$$Z: \quad \mathcal{C}ob_{d,d-1} \to \mathcal{V}ect(\mathbb{K})\,. \tag{2.88}$$

However, there is further essential structure on the two categories involved that is still neglected in our improved version. Namely, the tensor products on these categories are *symmetric*. For two-dimensional cobordisms, the symmetry is provided by a pair of *exchanging cylinders* (for an illustration, see the picture (3.5) below). This bears some similarity with the flip map (2.21) which we we already have used repeatedly, and which can be formulated for any pair of $\mathbb{K}$-vector spaces V and W:

$$\begin{aligned} \tau: \quad V\otimes W &\longrightarrow W\otimes V\,, \\ v\otimes w &\longmapsto w\otimes v\,. \end{aligned}$$

Rather than introducing separately the notion of a symmetric monoidal category, we directly discuss a more general class of categories that we will encounter later on anyhow. Let $(\mathcal{C},\otimes,a,l,r)$ be a monoidal category. Together with the tensor product $\otimes\colon \mathcal{C}\times\mathcal{C}\longrightarrow\mathcal{C}$ we also have an *opposite tensor product*, namely the functor

$$\otimes^{\text{opp}} := \otimes\circ\tau: \quad \mathcal{C}\times\mathcal{C}\longrightarrow\mathcal{C}$$

with

$$V\otimes^{\text{opp}} W := W\otimes V \qquad \text{and} \qquad f\otimes^{\text{opp}} g := g\otimes f\,.$$

This functor indeed furnishes another tensor product on $\mathcal{C}$: Given an associator a for $\otimes$, the family of morphisms

$$a^{\text{opp}}_{U,V,W} := a^{-1}_{W,V,U}$$

is an associator for the tensor product $\otimes^{\text{opp}}$. Left and right unit constraints for $\otimes^{\text{opp}}$ are obtained from those of $(\mathcal{C},\otimes)$ in a similar manner.

It is a natural question to ask whether we can compare the two tensor products $\otimes$ and $\otimes^{\text{opp}}$. This leads to the following structure.

DEFINITION 2.89.

(1) A *commutativity constraint* for a monoidal category $(\mathcal{C},\otimes)$ is the datum of a natural isomorphism

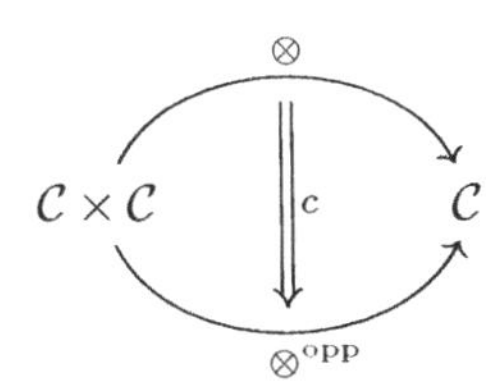

of functors from $\mathcal{C}\times\mathcal{C}$ to $\mathcal{C}$. Explicitly, we have an isomorphism

$$c_{V,W}: \quad V\otimes W \xrightarrow{\;\cong\;} W\otimes V$$

for any pair (V, W) of objects of $\mathcal{C}$, such that for every pair of morphisms $V \xrightarrow{f} V'$ and $W \xrightarrow{g} W'$ the diagram

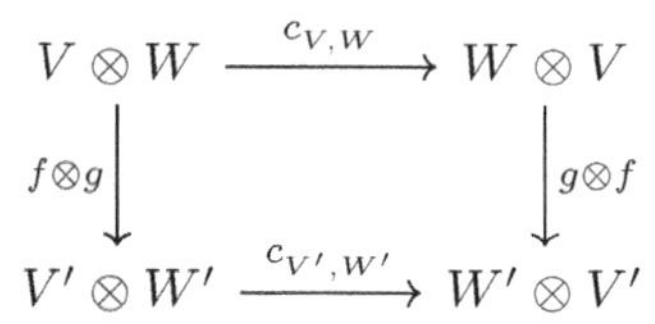

commutes.

(2) A *braiding* is a commutativity constraint such that for every triple of objects U, V and W the two diagrams

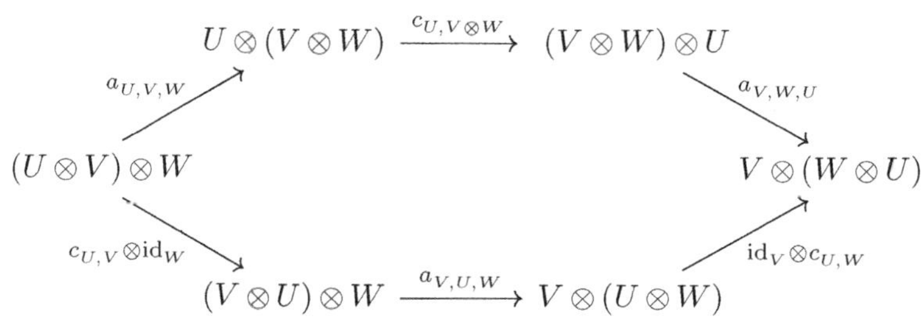

and

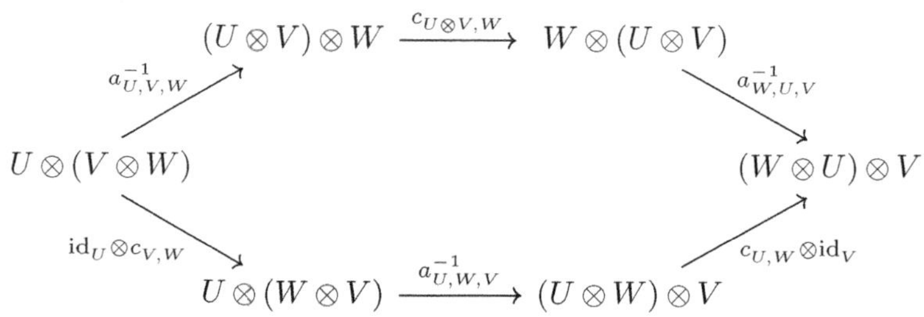

commute. These are called the *hexagon axioms*.

REMARK 2.90. For a *strict* monoidal category, the hexagon constraints simplify to

$$c_{U\otimes V,W} = (c_{U,W} \otimes \mathrm{id}_V) \circ (\mathrm{id}_U \otimes c_{V,W}) \tag{2.91}$$

and

$$c_{U,V\otimes W} = (\mathrm{id}_V \otimes c_{U,W}) \circ (c_{U,V} \otimes \mathrm{id}_W)\,.$$

DEFINITION 2.92.

(1) A *braided monoidal category* is a monoidal category together with the structure of a braiding.

(2) In case the family $c = (c_{U,V})$ of isomorphisms satisfies the identity $c_{U,V} = c_{V,U}^{-1}$, the braided monoidal category is called *symmetric*.

EXERCISE 2.93. Show that for any braiding $c = (c_{U,V})$ on a monoidal category $\mathcal{C}$, the family $(c_{V,U}^{-1})$ of isomorphisms is a braiding on $\mathcal{C}$ as well.

(Thus a symmetric monoidal category can be characterized as a monoidal category for which two a priori different braidings actually coincide.)

GRAPHICAL DESCRIPTION 2.94. In the graphical string calculus, we represent the braiding by a string diagram with an overcrossing of two strands, and its inverse by an analogous diagram with an undercrossing:

$c_{X,Y} \;=$ 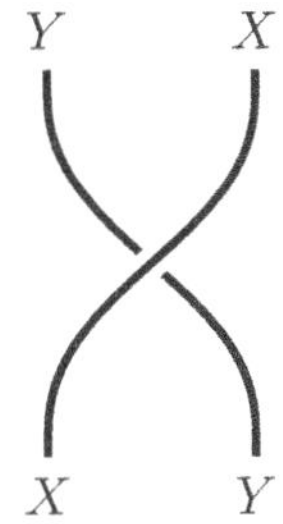

and $c^{-1}_{X,Y} \;=$ 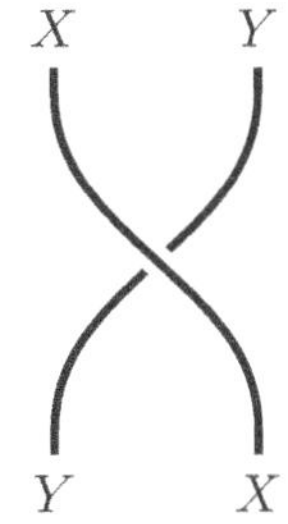

(For a symmetric category one has $c^{-1}_{X,Y} = c_{Y,X}$, so that there is no need to distinguish overcrossings from undercrossings.) Then, e.g., the first of the conditions (2.91) is the equality

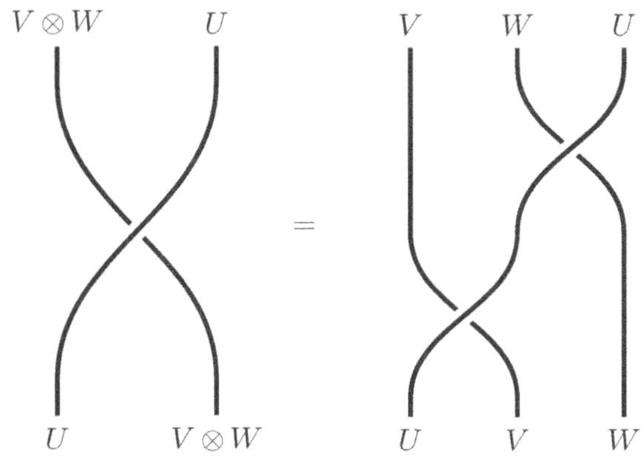

among string diagrams.

Note that the bialgebra axiom in Graphical Description 2.30 can now be given meaning in an arbitrary braided monoidal category, by using the braiding in place of the flip map of vector spaces. Pictorially:

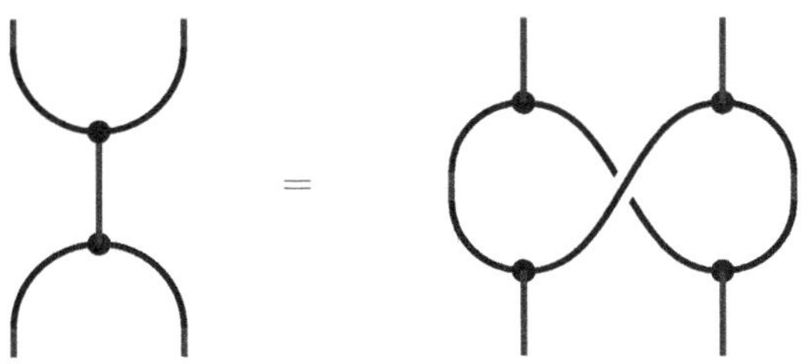

REMARK 2.95. With the chosen convention of depicting the braiding and its inverse as string diagrams with overcrossings and undercrossings, the observation in Remark 2.61 generalizes: Any closed graph drawn in $\mathbb{R}^2$ with overcrossings and undercrossings whose components are labeled by objects in a rigid braided category $\mathcal{C}$ can be evaluated to yield an element of $\mathrm{Hom}(\mathbf{1},\mathbf{1})$ and thus, in case the monoidal unit $\mathbf{1}$ is absolutely simple, a number. This is, for instance, of crucial importance for the construction of invariants of knots and links and of three-manifolds from representation-theoretic input data (compare Sections 5.6 and 6.4.3).

It is, however, worth being aware of the fact that a graph in $\mathbb{R}^2$ involving an overcrossing or undercrossing is actually not a proper graph in $\mathbb{R}^2$. Instead, it should be seen as the projection to $\mathbb{R}^2$ of a graph embedded in $\mathbb{R}^3$. In other words, the graphical string calculus for a braided category is really not two-dimensional, but *three*-dimensional.

REMARKS 2.96.

(1) The flip map

$$\tau: \quad V \otimes W \longrightarrow W \otimes V\,, \qquad v \otimes w \longmapsto w \otimes v$$

defines a symmetric braiding on the monoidal category $\mathcal{V}ect(\mathbb{K})$ of $\mathbb{K}$-vector spaces. It also induces a symmetric braiding on the category $\mathbb{K}[G]$-mod of $\mathbb{K}$-linear representations of a group G.

(2) There exist monoidal categories that do not admit a braiding. A simple example of a monoidal category not admitting any braiding is the category $\mathcal{V}ect_G(\mathbb{K})$ of G-graded vector spaces for G a nonabelian group. To see this, denote as in Example 2.47 by $\mathbb{K}_g$ the G-graded vector space that is given by the ground field concentrated in degree $g \in G$. Then we have

$$\mathbb{K}_g \otimes \mathbb{K}_h \cong \mathbb{K}_{gh} \qquad \text{and} \qquad \mathbb{K}_h \otimes \mathbb{K}_g \cong \mathbb{K}_{hg}$$

for $g, h \in G$. Thus whenever $gh \neq hg$, the two objects $\mathbb{K}_g \otimes \mathbb{K}_h$ and $\mathbb{K}_h \otimes \mathbb{K}_g$ are not isomorphic.

(3) The category $\mathcal{V}ect_G(\mathbb{K})$ admits the flip as a braiding if the group G is abelian. In the case of $G = \mathbb{Z}_2$, objects are $\mathbb{Z}_2$-graded vector spaces $V_0 \oplus V_1$. Besides the flip, there is another symmetric braiding c on this category: On homogeneous vector spaces, it equals the flip except for the case of two odd vector spaces, for which it is given by

$$c: \quad V_1 \otimes W_1 \longrightarrow W_1 \otimes V_1\,, \qquad v_1 \otimes w_1 \longmapsto -\, w_1 \otimes v_1\,.$$

The so obtained braided monoidal category is called the category of *super vector spaces*. (It is worth noting that in some treatments a fixed pivotal structure is understood as part of the structure of a super vector space, so that a trace is defined.)

This shows that a monoidal category can admit inequivalent braidings.

(4) As seen in Example 2.15(4), for any d, the monoidal category of d-dimensional cobordisms admits the disjoint union "$\sqcup$" as a tensor product. The cylinder over a disjoint union $M_1 \sqcup M_2$ of manifolds can be regarded as a cobordism $M_1 \sqcup M_2 \longrightarrow M_2 \sqcup M_1$ if the map that identifies the outgoing boundary of the cylinder with $M_2 \sqcup M_1$ is chosen accordingly. This prescription yields a symmetric braiding for the monoidal structure. For the two-dimensional case, this will be illustrated in Remark 3.6 below.

EXERCISE 2.97. Exhibit two different pivotal structures on the category of super vector spaces.

REMARK 2.98. For any positive integer n, we can consider the set of all braids on n strands – to be understood again as isotopy classes of vertically running braids in three-dimensional space, similarly as for the graphical calculus in a braided monoidal category, the only difference being that the braids are not labeled by an object. Under concatenation, this set forms a group, called the *braid group on n strands* and denoted by B_n. The group B_n is generated by the elements σ_i, for $i \in \{1, 2, \ldots, n-1\}$, which braid the ith strand over the $i+1$th one. Together with the relations

$$\sigma_i\,\sigma_j = \sigma_j\,\sigma_i \quad \text{for } |i-j| \geq 2$$

and

$$\sigma_i\,\sigma_{i+1}\,\sigma_i = \sigma_{i+1}\,\sigma_i\,\sigma_{i+1} \quad \text{for } 1 \leq i \leq n-2\,,$$

this provides a description of the group B_n in terms of generators and relations.

Adding $\sigma_i^2 = e$, for every i, as further relations one obtains the *symmetric group* S_n. Thus there is a surjective group homomorphism $B_n \longrightarrow S_n$. While the symmetric group S_n has finite order $n!$, the order of B_n is infinite for all $n \geq 2$.

As is easily seen via the graphical calculus, for a braided monoidal category $\mathcal{C}$ the nth tensor power $X^{\otimes n}$ of any object $X \in \mathcal{C}$ comes with a canonical B_n-action, namely the one sending the generator σ_i to the automorphism that is induced by the braiding isomorphism $c_{X,X}\colon X \otimes X \longrightarrow X \otimes X$ inserted at the appropriate place, i.e., to

$$X^{\otimes(i-1)} \otimes X \otimes X \otimes X^{\otimes(n-i-1)} \xrightarrow{\mathrm{id}_{X^{\otimes(i-1)}} \otimes c_{X,X} \otimes \mathrm{id}_{X^{\otimes(n-i-1)}}} X^{\otimes(i-1)} \otimes X \otimes X \otimes X^{\otimes(n-i-1)}.$$

This statement is often paraphrased by saying that "X together with $c_{X,X}$ solves the Yang-Baxter equation"; we refer to Chapter VIII of [**Kas**] for details.

Having introduced the additional structure of a braiding on categories, we again need corresponding types of functors and natural transformations that obey appropriate compatibility conditions:

DEFINITION 2.99.

(1) A monoidal functor $(F, \varphi_0, \varphi_2)$ from a braided monoidal category $\mathcal{C}$ to a braided monoidal category $\mathcal{D}$ is called a *braided monoidal functor* (or braided functor) if for any pair of objects (V, V') of $\mathcal{C}$ the square

$$\begin{array}{ccc} F(V) \otimes F(V') & \xrightarrow{\varphi_2(V,V')} & F(V \otimes V') \\ \Big\downarrow{\scriptstyle c^{\mathcal{D}}_{F(V),F(V')}} & & \Big\downarrow{\scriptstyle F(c^{\mathcal{C}}_{V,V'})} \\ F(V') \otimes F(V) & \xrightarrow{\varphi_2(V',V)} & F(V' \otimes V) \end{array}$$

in $\mathcal{D}$ commutes.

If the braided monoidal categories $\mathcal{C}$ and $\mathcal{D}$ have the property of being symmetric, a braided monoidal functor between them is also called a *symmetric monoidal functor*.

(2) As braided monoidal natural transformations, we take all monoidal natural transformations.

The natural monoidal structure (2.42) that is carried by the composite $G \circ F$ of the functors that underlie two braided monoidal functors is again braided.

We are now finally ready to state the complete definition of a topological field theory:

DEFINITION 2.100. Let $\mathbb{K}$ be an algebraically closed field. A *topological field theory* of dimension d is a symmetric monoidal functor

$$Z: \quad \mathcal{C}ob_{d,d-1} \to \mathcal{V}ect(\mathbb{K}).$$

REMARK 2.101. In the definition one may replace the category $\mathcal{V}ect(\mathbb{K})$ by any symmetric monoidal category. Interesting examples include, e.g., categories of complexes of vector spaces.

Let us spell out the contents of Definition 2.100 in more detail: A topological field theory Z of dimension d is given by the following data:

(1) For every closed oriented manifold M of dimension $d-1$, a $\mathbb{K}$-vector space $Z(M)$.

(2) For every oriented cobordism B from a $(d-1)$-manifold M to a $(d-1)$-manifold N, a $\mathbb{K}$-linear map $Z(B)\colon Z(M) \longrightarrow Z(N)$.

   Cobordisms that coincide up to a diffeomorphism relative to the boundary (that is, up to a diffeomorphism that is the identity on ∂B) give the same linear map.

(3) An isomorphism

$$Z(\emptyset) \cong \mathbb{K}$$

(with the empty set regarded as a $(d-1)$-manifold) and a collection of isomorphisms

$$Z(M \sqcup N) \cong Z(M) \otimes Z(N),$$

one for each pair of $(d-1)$-manifolds.

By the functoriality of Z, the gluing of cobordisms is mapped to the composition of linear maps. Moreover, these data are required to satisfy a number of natural coherence properties which can be obtained by writing out the structure implied by the fact that Z is a symmetric monoidal functor.

A *closed* oriented manifold M of dimension d can be regarded as a cobordism from the empty $(d-1)$-manifold to itself, $M\colon \emptyset \longrightarrow \emptyset$. Thus we have

$$Z(M): \quad \mathbb{K} \cong Z(\emptyset) \longrightarrow Z(\emptyset) \cong \mathbb{K}$$

so that $Z(M) \in \mathrm{Hom}_{\mathbb{K}}(\mathbb{K},\mathbb{K}) \cong \mathbb{K}$, i.e., $Z(M)$ is a number. This number depends on the manifold M only up to diffeomorphisms; it thus constitutes the value of an *invariant* assigned to every closed oriented manifold of dimension d. The quantities that the topological field theory assigns to manifolds of dimension $d-1$ are needed to build up arbitrary d-dimensional manifolds from simple pieces. That the topological field theory includes such assignments is thus a necessary ingredient for the locality of the theory.

Topological field theories with a fixed target category, say $\mathcal{V}ect(\mathbb{K})$, and fixed dimension d form themselves the objects of a category. We denote this category by $[\mathcal{C}ob_{d,d-1}, \mathcal{V}ect(\mathbb{K})]_{\otimes}$. Its morphisms are monoidal natural transformations.

PROPOSITION 2.102. Every morphism between d-dimensional topological field theories is invertible, i.e., the category of d-dimensional topological field theories is in fact a groupoid.

EXERCISE 2.103. Using Exercise 2.63, prove Proposition 2.102.
For a hint, the reader may consult Theorem 2.5 of [**Pst**].

REMARK 2.104. Let Z be a d-dimensional topological field theory. Let Σ be a closed oriented manifold of dimension $d-1$ and $f\colon \Sigma \longrightarrow \Sigma$ an orientation preserving diffeomorphism. Then f gives rise to a cobordism $\Sigma \longrightarrow \Sigma$, given by the cylinder $\Sigma \times [0,1]$ together with the map

$$\overline{\Sigma} \sqcup \Sigma \xrightarrow{\mathrm{id}\times f} \overline{\Sigma} \sqcup \Sigma = \partial(\Sigma \times [0,1])\,.$$

The equivalence class of this cobordism only depends on the isotopy class of the orientation preserving diffeomorphism f. We denote the resulting morphism in the cobordism category by $C_f\colon \Sigma \longrightarrow \Sigma$. Such a cobordism (class) is often referred to as a *mapping cylinder*.

This construction can be used to extract representations of mapping class groups from a topological field theory: The *mapping class group* $\mathrm{Map}(\Sigma)$ of Σ is by definition the group of isotopy classes of orientation preserving diffeomorphisms $\Sigma \longrightarrow \Sigma$. By setting

$$\mathrm{Map}(\Sigma) \to \mathrm{Aut}(Z(\Sigma))\,, \quad [f] \longmapsto Z(C_f)$$

one then obtains an action of the mapping class group of Σ on $Z(\Sigma)$.

The reader is encouraged to check the statements in this remark in detail.

We now indicate how to describe a braiding on the monoidal category of modules over a bialgebra.

DEFINITION 2.105.

(1) Let H be a bialgebra. The structure of a *quasi-cocommutative bialgebra* on H is the choice of an invertible element R in the algebra $H \otimes H$ such that

$$\Delta^{\mathrm{copp}}(x)\, R = R\, \Delta(x) \quad \text{for all } x \in H\,.$$

The element R is called a *universal R-matrix.*

A quasi-cocommutative Hopf algebra is a Hopf algebra together with the choice of a universal R-matrix.

(2) A quasi-cocommutative bialgebra H is called *quasi-triangular* if its universal R-matrix obeys the relations

$$(\Delta \otimes \mathrm{id}_H)(R) = R_{13}\, R_{23} \qquad \text{and} \qquad (\mathrm{id}_H \otimes \Delta)(R) = R_{13}\, R_{12}$$

in $H \otimes H \otimes H$, where $R_{12} := R \otimes 1$, $R_{23} := 1 \otimes R$ and

$$R_{13} := (\tau_{H,H} \otimes \mathrm{id}_H)(1 \otimes R)\,.$$

(3) A morphism $f\colon (H,R) \longrightarrow (H',R')$ of quasi-triangular Hopf algebras is a morphism $f\colon H \longrightarrow H'$ of Hopf algebras such that $R' = (f \otimes f)(R)$.

EXAMPLE 2.106. Consider the cocommmutative Hopf algebra $\mathbb{K}[\mathbb{Z}_2]$ for $\mathbb{K}$ a field of characteristic different from 2. Write $\mathbb{Z}_2$ as $\{e,g\}$. Then

$$R := \frac{1}{2}\,(\beta_e \otimes \beta_e + \beta_e \otimes \beta_g + \beta_g \otimes \beta_e - \beta_g \otimes \beta_g)$$

is a universal R-matrix for $\mathbb{K}[\mathbb{Z}_2]$.

The four-dimensional Taft Hopf algebra H_2 from Example 2.81 admits a one-parameter family of R-matrices [**Kas**, p. 174].

EXERCISE 2.107. Let A be a bialgebra. The monoidal category A-mod is braided if and only if A is quasi-triangular. The two structures are in bijection. To prove this, show:

(1) If A is quasi-triangular with R-matrix R, then for any pair U, V of left A-modules, the linear map

$$\begin{aligned} c^R_{U,V} : \quad & U \otimes V \longrightarrow V \otimes U\,, \\ & u \otimes v \longmapsto \tau_{U,V}(R\,(u \otimes v)) = R_2\, v \otimes R_1\, u \end{aligned}$$

defines a braiding on A-mod.

Here we use the abbreviation $R = R_1 \otimes R_2$ for $R \in A \otimes A$, in a similar spirit as in the Sweedler notation for the coproduct.

(2) Conversely, suppose that the category A-mod is endowed with a braiding. Then the element

$$R := \tau_{A,A}(c_{A,A}(1_A \otimes 1_A)) \;\in A \otimes A$$

is an R-matrix for A and the braiding defined by this R-matrix is the original braiding.

CHAPTER 3

Topological field theories in dimension 1 and 2

In Definition 2.100 we have introduced the notion of a topological field theory in any dimension d. We now proceed by investigating topological field theories for small values of d. In dimension 1 or 2, classical results allow us to a describe a topological field theory in terms of algebraic structures that are sufficiently simple such that they are also studied for reasons independent of TFT.

3.1. One-dimensional topological field theories

By definition, a *one*-dimensional topological field theory is a symmetric monoidal functor

$$Z: \quad \mathcal{C}ob_{1,0} \longrightarrow \mathcal{V}ect(\mathbb{K}).$$

The functor Z assigns a vector space $Z(M)$ to every closed oriented 0-manifold M, i.e., to every finite set of oriented points. Since Z is a monoidal functor, it suffices to know its values $Z(\bullet,+)$ and $Z(\bullet,-)$ on the positively and negatively oriented point. Note that according to Remark 2.64 these vector spaces are finite-dimensional and are dual to each other. Setting $V := Z(\bullet,+)$ it follows that on any 0-manifold M the value of Z is

$$Z(M) \cong \Big(\bigotimes_{x\in M_+} V\Big) \otimes \Big(\bigotimes_{y\in M_-} V^*\Big),$$

with $M_\pm \subset M$ the subsets of positively and of negatively oriented points, respectively.

We must also give the value of Z on any one-manifold B with boundary. Since Z is a monoidal functor, it suffices to specify the linear map $Z(B)$ for the case that B is *connected.* There are precisely two such one-manifolds B up to diffeomorphism: B is diffeomorphic either to a closed interval $[0,1]$ or to a circle S^1. There are then five cases to consider, which differ by the way we interpret the one-dimensional manifold B with boundary as a cobordism:

(1) Suppose that $B=[0,1]$, regarded as a cobordism from the positive point $(\bullet,+)$ to itself.

Then $Z(B)$ is the identity map $\mathrm{id}_V : V \rightarrow V$, since any functor preserves identities.

(2) Suppose that $B=[0,1]$, regarded as a cobordism from the negative point $(\bullet,-)$ to itself.

Then $Z(B)$ is the identity map $\mathrm{id}_{V^*} : V^* \rightarrow V^*$, again because any functor preserves identities.

(3) Suppose that $B=[0,1]$, regarded as a cobordism from the disjoint union $(\bullet,+) \sqcup (\bullet,-)$ to the empty set.

Then $Z(B)$ is a linear map from $V\otimes V^*$ into the ground field $\mathbb{K}$, namely the evaluation map $(v,\lambda) \mapsto \lambda(v)$. (This follows, e.g., from Exercise 2.63.)

Since the functor Z is not just monoidal, but even symmetric monoidal, the evaluation of Z on the cobordism from $(\bullet,-)\sqcup(\bullet,+)$ to the empty set can be described analogously.

(4) Suppose that $B=[0,1]$, regarded as a cobordism from the empty set to the disjoint union $(\bullet,+)\sqcup(\bullet,-)$.

Then $Z(B)$ is a linear map from $\mathbb{K}$ to the vector space

$$Z((\bullet,+)\sqcup(\bullet,-))\cong V\otimes V^*\cong \mathrm{End}(V)\,.$$

This linear map is given by the coevalution $x\mapsto x\,\mathrm{id}_V$.

Again, we can exchange the order of the objects, thanks to symmetry of $\mathcal{C}ob_{1,0}$.

(5) Suppose that $B=S^1$, which is to be regarded as a cobordism from the empty set to itself.

Then $Z(B)$ is a linear map from the field $\mathbb{K}$ to itself, which can thus be identified with an element of $\mathbb{K}$. To determine this element, decompose the circle $S^1\cong\{z\in\mathbb{C}\mid |z|=1\}$ into two intervals

$$S^1_-:=\{z\in S^1\mid \mathrm{Im}\,(z)\leq 0\}\qquad\text{and}\qquad S^1_+:=\{z\in S^1\mid \mathrm{Im}\,(z)\geq 0\}$$

with intersection

$$S^1_-\cap S^1_+=\{1\,,-1\}\subset S^1.$$

It follows that $Z(S^1)$ is the composition

$$\mathbb{K}\cong Z(\emptyset)\xrightarrow{Z(S^1_-)} Z(\{1,-1\})\xrightarrow{Z(S^1_+)} Z(\emptyset)\cong\mathbb{K}$$

of linear maps. These maps are the ones described in the cases (3) and (4) above. Thus after choosing a basis $(v_i)_{i\in I}$ of V and a dual basis $(\beta^i)_{i\in I}$ of V^*, we get the map

$$\begin{aligned}\mathbb{K}\cong Z(\emptyset)&\longrightarrow Z(\{\pm1\})\cong V\otimes V^*\longrightarrow Z(\emptyset)\cong\mathbb{K}\,,\\ \lambda\mapsto\ &\quad\lambda\textstyle\sum_i\beta^i\otimes v_i\qquad\mapsto\lambda\sum_i\beta^i(v_i)=\lambda\,\dim V.\end{aligned}$$

We thus conclude that the number $Z(S^1)$ is the dimension of the vector space $V=Z(\bullet,+)$.

Note that via the functor Z, the category $\mathcal{C}ob_{1,0}$ – one of the simplest geometric systems – has led us to one of the most fundamental algebraic systems: a finite-dimensional vector space with its dual and its tensor powers, and with the dimension of the vector space as an invariant. In fact, a one-dimensional topological field theory leads *precisely* to these data, in the sense that the following classification is valid:

THEOREM 3.1. Evaluation on the positively oriented point yields an equivalence

$$[\mathcal{C}ob_{1,0},\mathcal{V}ect(\mathbb{K})]_\otimes\xrightarrow{\simeq}\mathcal{V}ect(\mathbb{K})^\times$$

from the groupoid of one-dimensional topological field theories to the maximal subgroupoid $\mathcal{V}ect(\mathbb{K})^\times\subset\mathcal{V}ect(\mathbb{K})$ of the category of finite-dimensional vector spaces.

That evaluation on the positively oriented point provides a functor to $\mathcal{V}ect(\mathbb{K})^\times$ is clear from the discussion above and Proposition 2.102. Theorem 3.1 makes the stronger statement that from an oriented one-dimensional topological field theory we cannot only extract a finite-dimensional vector space, but that this vector space is indeed equivalent to giving an oriented one-dimensional topological field theory. This is strongly suggested by the fact that, for any one-dimensional topological field theory Z, all morphisms in $\mathcal{C}ob_{1,0}$ (we have already listed all of them) give us either the identity of $Z(\bullet, +)$ or the (co)evaluation of the dualizable vector space $Z(\bullet, +)$. This can be turned into a formal proof; see Section 2.5 of [**Saf**] for more details.

REMARK 3.2. In the literature, topological field theories (or TFTs) are often alternatively referred to as *topological quantum field theories* (TQFTs). A justification of this terminology is to realize that in physics, one-dimensional field theories are frequently related to mechanical systems, so that it is natural to think of a one-dimensional topological field theory in terms of an, admittedly very simple, quantum mechanical system.

To corroborate this point of view, we reason as follows. To a point, the one-dimensional topological field theory assigns a vector space V. We interpret this vector space as the space of states of a quantum mechanical system. (The reader might expect that we actually deal with a Hilbert space. But as our framework is algebraic, it does not allow us to anticipate a Hilbert space structure and unitary operators.) To a closed interval, the topological field theory assigns a linear endomorphism of V; we interpret this endomorphism as the time evolution operator U of the quantum mechanical system. Now by construction U is actually the identity endomorphism of V, so it amounts to trivial time evolution and hence to a Hamilton operator $H=0$. This behavior might appear unattractive, but it is in fact perfectly appropriate. Indeed, in applications the vector space V models, quite generally, a space of ground states rather than the full state space of a system. Since each ground state has the same energy eigenvalue, the time evolution on the space of ground states is a scalar multiple of the identity, and the value of the scalar does not have any intrinsic meaning. The only invariant of the system is the degeneracy $\dim V$ of the space of ground states.

This idea has important ramifications for higher-dimensional topological field theories. It can be made very explicit for three-dimensional theories, e.g., in the framework of *Kitaev models*. For these, the vector space assigned by the TFT to a surface is the space of ground states of a set of commuting idempotents [**BuMCA**].

In this context, the linear map associated to the coevaluation in the cobordism category describes an "exotic time evolution" not customarily considered in quantum physics: Starting out with a trivial system with a one-dimensional state space, a nontrivial system consisting of two coupled subsystems is created. Analogously, the evaluation describes the simultaneous disappearance of two quantum mechanical systems. Since this exotic behavior arises from the presence of dualities, it is ultimately responsible for the fact that the theory describes systems with finite-dimensional state spaces, compare Lemma 2.62.

REMARK 3.3. In Theorem 3.1 we may replace the symmetric monoidal category of vector spaces over a field with any other symmetric monoidal category $\mathcal{C}$. We then obtain an equivalence from one-dimensional topological field theories with values in $\mathcal{C}$ to the groupoid of dualizable objects in $\mathcal{C}$.

This way we find, for example, that a one-dimensional topological field theory with values in the category R-mod of left modules over a commutative ring R can equivalently be described as a dualizable R-module.

EXERCISE 3.4. Show that a module over a commutative ring R is dualizable if and only if it is finitely generated and projective.

Hint: To show that a finitely generated projective module is dualizable, one can first prove that dualizable objects are preserved under direct sums and retracts. Knowing that any finitely generated projective R-module is a retract of $R^{\oplus n}$ for some $n \geq 0$, one can then reduce the statement to the dualizability of R.

For the converse, one can first use the Yoneda lemma to see that the (left and right) dual module $M^\vee$ of some R-module, provided it exists, is isomorphic to the R-module $\mathrm{Hom}_R(M, R)$. Finite generation and projectiveness then follow from the snake relations.

3.2. Topological field theories in dimension 2

Not surprisingly, characterizing two-dimensional topological field theories is richer than the one-dimensional case. However also in dimension 2, topological field theories can be described in terms of customary algebraic data.

A two-dimensional topological field theory Z assigns a finite-dimensional vector space $Z(M)$ to every closed oriented one-manifold M. Any such manifold is diffeomorphic to a finite disjoint union of circles, $M \cong (S^1)^{\sqcup n}$ for some $n \geq 0$. Since Z is monoidal, we have

$$Z(M) \cong A^{\otimes n} \qquad \text{with} \qquad A := Z(S^1)$$

a finite-dimensional vector space.

As mentioned in Example 1.39, the monoidal category $\mathcal{C}ob_{2,1}$ is generated under composition and disjoint union by the following six cobordisms: the cap (or disk, with the boundary circle being outgoing), the trinion (also called pair of pants), the cylinder, the opposite trinion which has two outgoing circles, the opposite cap (a disk with one incoming circle), and a pair of exchanging cylinders. In pictures,

(3.5)

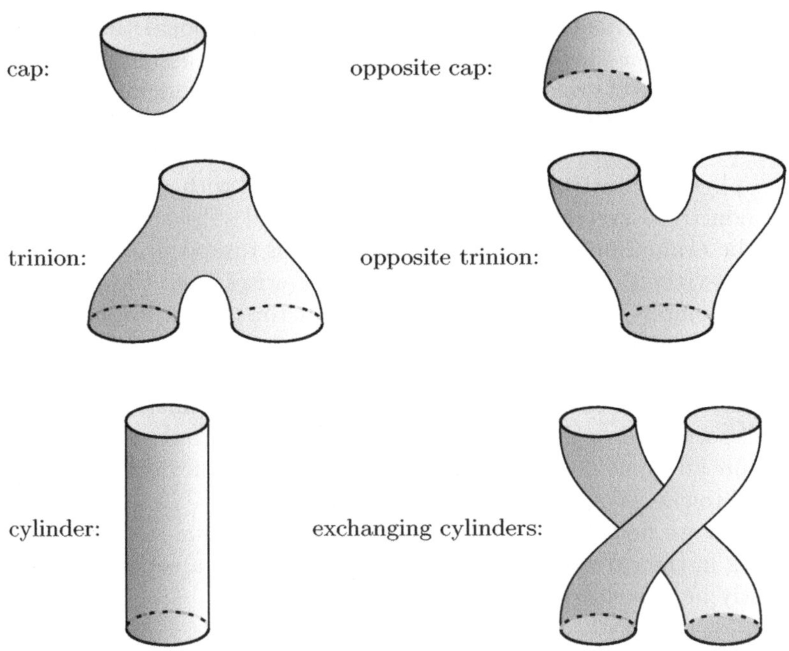

REMARK 3.6. Concerning the picture for the exchanging cylinders a word of warning is in order: It is an attempt to visualize a cobordism $S^1 \sqcup S^1 \to S^1 \sqcup S^1$, and does *not* show a two-manifold embedded in $\mathbb{R}^3$. Indeed, according to Definition 1.36 any such cobordism is represented by the two-manifold $B := (S^1 \sqcup S^1) \times [0,1]$ – i.e., the disjoint union of two cylinders over the circle – together with two maps $\Phi_{1,2}\colon S^1 \sqcup S^1 \to B$. In the case of the exchanging cylinders, these maps are as follows. Denote points in the first summand of $S^1 \sqcup S^1$ by $(z,1)$, points in the second summand by $(z,2)$, with $z \in S^1$, and points in B by triples (z,t,i) with $z \in S^1$, $t \in [0,1]$ and $i \in \{1,2\}$. Then we have $\Phi_1((z,i)) = (\overline{z},0,i)$ and $\Phi_2((z,i)) = (z,1,3-i)$. Thus the two circles in the left-hand part of the picture show the first summand in $S^1 \sqcup S^1$, while those on the right-hand part show the second summand. In the visualization, this leads to an apparent crossing of the two cylinders.

Applying the functor Z to the generating cobordisms yields the following linear maps:

cap	$\eta:$	$\mathbb{K} \to A\,,$
trinion	$\mu:$	$A \otimes A \to A\,,$
cylinder	$\mathrm{id}_A:$	$A \to A\,,$
opposite trinion	$\Delta:$	$A \to A \otimes A\,,$
opposite cap	$\varepsilon:$	$A \to \mathbb{K}\,,$
exchanging cylinders	$\tau:$	$A \otimes A \to A \otimes A\,.$

The relations between the cobordisms just listed imply that A has the structure of a *commutative Frobenius algebra*. (For a careful exposition of $\mathcal{C}ob_{2,1}$ in terms of the generators (3.5) and their relations see [**Ab**] and Chapter 1.4 of [**Koc**].) This insight will in fact lead to a classification of two-dimensional topological field theories by commutative Frobenius algebras. We are going to introduce the notion of a Frobenius algebra in the next subsection. Frobenius algebras also occur in other areas; for instance, the cohomology ring of any compact oriented manifold has a natural structure of a Frobenius algebra.

3.3. Frobenius algebras

We have seen that the evaluation of a two-dimensional topological field theory on the trinion gives a map $\mu\colon A \otimes A \to A$. It is easily verified that this map is associative and unital, i.e., it endows A with the structure of an (associative, unital) algebra. Similarly, the linear map $\Delta\colon A \to A \otimes A$ obtained by evaluation on the opposite trinion, which has domain and codomain exchanged as compared to the product, satisfies a dual version of associativity and unitality. This turns A into a (coassociative, counital) coalgebra.

Also note that via multiplication from the left and right, respectively, A is, as any associative algebra, a bimodule over itself. Similarly, $A \otimes A$ becomes an A-bimodule via left multiplication on the first tensor factor and right multiplication on the second tensor factor, respectively.

DEFINITION 3.7. Let (A, μ, η) be a unital associative algebra over a field $\mathbb{K}$, and require that the unit of A is not zero.

(1) A *(Δ, ε)-Frobenius structure* on A is the structure of a coassociative, counital coalgebra (A, Δ, ε) such that $\Delta\colon A \longrightarrow A \otimes A$ is a morphism of A-bimodules.
(2) A *κ-Frobenius structure* on A is a bilinear pairing $\kappa \in \mathrm{Hom}_{\mathbb{K}}(A \otimes A, \mathbb{K})$ that is invariant (or associative), i.e., satisfies

$$\kappa \circ (\mu \otimes \mathrm{id}_A) = \kappa \circ (\mathrm{id}_A \otimes \mu)\,,$$

and that is nondegenerate.

GRAPHICAL DESCRIPTION 3.8. In terms of string diagrams, the bimodule condition in the definition of a (Δ, ε)-Frobenius structure reads

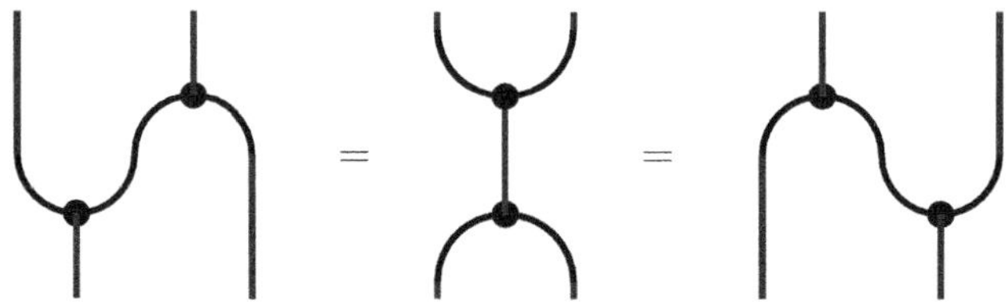

where all lines are labeled by A.

EXERCISE 3.9. Show – for instance graphically – that any of the two equalities in the definition of a (Δ, ε)-Frobenius structure follows from the other when combined with the algebra and coalgebra relations.

The notions of a (Δ, ε)-Frobenius structure and of a κ-Frobenius structure on an algebra (A, μ, η) are equivalent – accordingly, an algebra equipped with any of these structures is just called a *Frobenius algebra.* More concretely, we have:

(1) If $(A, \mu, \eta, \Delta, \varepsilon)$ is an algebra with a (Δ, ε)-Frobenius structure, then $(A, \mu, \eta, \kappa_\varepsilon)$ with

$$\kappa_\varepsilon := \varepsilon \circ \mu$$

is an algebra with κ-Frobenius structure.
(2) Let (A, m, η, κ) be an algebra with a κ-Frobenius structure. Owing to the nondegeneracy of the bilinear form κ, there exists an isomorphism $\Phi_\kappa \in \mathrm{Hom}(A, A^*)$ of vector spaces. Introduce the morphisms

$$\Delta_\kappa := (\mathrm{id}_A \otimes \mu) \circ (\mathrm{id}_A \otimes \Phi_\kappa^{-1} \otimes \mathrm{id}_A) \circ (b_A \otimes \mathrm{id}_A) \qquad \text{and}$$
$$\varepsilon_\kappa := \kappa \circ (\mathrm{id}_A \otimes \eta)\ .$$

Then $(A, \mu, \eta, \Delta_\kappa, \varepsilon_\kappa)$ is an algebra with (Δ, ε)-Frobenius structure.

If the pairing κ is symmetric, then A is called a *symmetric Frobenius algebra.*

DEFINITION 3.10.

(1) A *morphism* $\phi\colon A \longrightarrow B$ of Frobenius algebras (treated as (Δ, ε)-Frobenius algebras) is a linear map that is an algebra morphism and a coalgebra morphism.
(2) We denote by $\mathcal{F}rob$ and $cm\mathcal{F}rob$ the *category of Frobenius algebras* and the one of *commutative Frobenius algebras* (over a fixed field), respectively.

LEMMA 3.11. The categories $\mathcal{F}rob$ and $cm\mathcal{F}rob$ are groupoids.

PROOF. Since $cm\mathcal{F}rob \subset \mathcal{F}rob$ is a full subcategory of $\mathcal{F}rob$, it suffices to prove the statement for $\mathcal{F}rob$. To do so, we first observe that the nondegeneracy of the pairing $\kappa_A\colon A \otimes A \to \mathbb{K}$ of a Frobenius algebra A implies that the null space $N(\varepsilon_A) := \{a \in A \,|\, \varepsilon_A(a) = 0\}$ does not contain any nontrivial (i.e., different from 0 or A) left or right ideals of A. To see this, consider for $a \in N(\varepsilon_A)$ the left ideal $I(a)$ generated by a. If $I(a) \subseteq N(\varepsilon_A)$, then for every $b \in A$ we have $ba \in N(\varepsilon_A)$ so that $\varepsilon_A(ba) = 0$, thus contradicting the nondegeneracy of ε_A unless $a = 0$. Since $I(a)$ is the smallest left ideal containing a, we conclude that for $a \neq 0$ the null space $N(\varepsilon_A)$ cannot contain any left ideal containing a. Analogously, $N(\varepsilon_A)$ cannot contain any right ideal, nor any two-sided ideal, except for the null ideal.

Consider now a morphism $\phi\colon A \to B$ of Frobenius algebras and assume that $a \in \ker\phi$. Then we have $\varepsilon_A(a) = \varepsilon_B(\phi(a)) = 0$, and hence $\ker\phi \subseteq N(\varepsilon_A) = \{0\}$. This proves that ϕ is injective.

Next observe that A^* and B^* endowed with the dual structure maps are again Frobenius algebras. Then $\phi^*\colon B^* \longrightarrow A^*$ is a map of Frobenius algebras as well, and therefore injective by the previous argument. This shows that ϕ is surjective. □

REMARK 3.12. There are finite-dimensional algebras which do not admit any structure of a Frobenius algebra [**SkY1**]. Such an algebra is necessarily nonsemisimple – in the case of algebras over the complex numbers, it is not isomorphic to a direct sum of full matrix algebras.

For the interested reader we mention that such an algebra can still be *self-injective*, also called *quasi-Frobenius*, which means that it is injective as a module over itself or, equivalently, that all projective modules are injective and vice versa. In this class of algebras, the smallest algebra that does not admit a Frobenius structure is nine-dimensional [**SkY2**, p. 391]. Self-injective algebras are Morita equivalent to Frobenius algebras, which in particular implies that the property of admitting a Frobenius structure is not preserved under Morita equivalence. For further details, see, e.g., Chapter IV.3 of [**SkY2**].

REMARK 3.13. Frobenius algebras cannot only be defined in the symmetric monoidal category of vector spaces, but in fact in any monoidal category [**FucS**] – it is not needed that the tensor product is symmetric. In particular, the notions of an algebra and of a coalgebra, and hence also the one of a (Δ, ε)-Frobenius algebra, generalizes in a straightforward manner. The required compatibility between product and coproduct as displayed in Graphical Description 3.8 can then directly be interpreted in the string calculus of the ambient monoidal category. Frobenius algebras in not necessarily symmetric monoidal categories play an important role in the construction of consistent systems of correlators in two-dimensional conformal field theories [**RuFFS**, **FS3**, **FSWY**]. For the Frobenius algebras occurring in this application, the κ-Frobenius structure is a symmetric pairing.

3.4. Classifying two-dimensional topological field theories

One can classify all relations between the generators of $\mathcal{C}ob_{2,1}$ that were shown in Section 3.2; see [**Koc**, 1.4.24 – 1.4.28]. The relations can be summarized as the statement that the category $\mathcal{C}ob_{2,1}$ is the free symmetric monoidal category on a commutative Frobenius object. This leads to the following result:

THEOREM 3.14. Evaluation on the circle yields an equivalence

$$[\mathcal{C}ob_{2,1}, \mathcal{V}ect(\mathbb{K})]_\otimes \simeq cm\mathcal{F}rob$$

from the groupoid of two-dimensional topological field theories to the groupoid of commutative Frobenius algebras.

REMARK 3.15. The phrase "evaluation on the circle" used here is a common slight abuse of language. The value that a two-dimensional topological field theory attains on the circle is clearly just a vector space. But it is understood that we also account for the evaluation on other genus-zero surfaces so as to obtain the Frobenius structure, as explained in Section 3.2.

According to Theorem 3.14, the description of $\mathcal{C}ob_{2,1}$ in terms of generators and relations tells us how to obtain a two-dimensional topological field theory from a commutative Frobenius algebra. We illustrate this result in the following exercise:

EXERCISE 3.16. Let A be a commutative complex Frobenius algebra that is in addition *semisimple*, i.e., is a direct sum of full matrix algebras. Then we can find a basis of commuting idempotents $e_1, e_2, \ldots, e_n \in A$, satisfying $e_i e_i = e_i$ and $e_i e_j = 0$ for $i \neq j$.

The structure of a Frobenius algebra is then determined by the nonzero complex numbers $\varepsilon(e_i) =: \lambda_i$ for $i = 1, 2, \ldots, \dim A$. For a commutative semisimple algebra, we thus have a whole family of different Frobenius *structures*.

(1) Show that the natural identification $\Phi^{-1}\colon A^\vee \longrightarrow A$ resulting from the Frobenius algebra structure identifies e_i^* with $\lambda_i^{-1} e_i$.
(2) Show that the two-holed torus maps to the endomorphism of A that is given by $e_i \mapsto \lambda_i^{-1} e_i$.
(3) Use a decomposition of a cobordism into pairs of pants and caps to show that a closed surface Σ_g of genus g maps to the complex number

$$Z(\Sigma_g) = \sum_{i=1}^{\dim A} \lambda_i^{1-g}. \tag{3.17}$$

CHAPTER 4

Dijkgraaf-Witten topological field theories

In this chapter we present the construction of a topological field theory that is based on the gauge theory for a finite group G as discussed in Section 2.1. Owing to its origin in [**DiW**], the so obtained topological field theories are known as *Dijkgraaf-Witten theories*. These theories were further elaborated and developed in, e.g., [**FrQ, Mor3, ScW**]. While our interest is primarily in three-dimensional models, Dijkgraaf-Witten theories provide examples of topological field theories in any dimension.

4.1. Finite path integrals and groupoid cardinality

Our aim is to associate to any finite group G a geometric construction of a topological field theory functor, to be denoted by Z_G. To achieve this, we use linearization techniques for essentially finite groupoids, following in particular the exposition in [**BaeHW**] and [**Mor1**]. In the present section we focus on the invariants that Z_G assigns to closed manifolds of top dimension.

Our starting point is to take up the idea, already put forward in the heuristic ansatz (2.3), that to a closed oriented d-dimensional manifold we can assign a number $Z_G(M)$ by means of a "path integral":

$$Z_G(M) = \text{“} \int_{\mathcal{B}un_G(M)} \mathrm{d}\Phi \; \mathrm{e}^{\mathrm{i}\, S[\Phi]} \text{”}. \tag{4.1}$$

Here $\mathcal{B}un_G(M)$ is the groupoid of G-bundles over M, as introduced in Definition 1.30. Let us emphasize again that a priori this path integral is merely a heuristic expression. The challenge is to make precise sense of it in the situation at hand. We have already made a first step toward achieving this goal by realizing that the field configurations of a quantum field theory form a category – in the present context, the groupoid $\mathcal{B}un_G(M)$. For achieving it in full it will be crucial to organize also other pertinent quantities in categorical terms.

In the case of a path integral (4.1) with vanishing action, $S[\Phi]=0$, the challenge amounts to finding a suitable measure of "size" or "volume" for the essentially finite groupoid $\mathcal{B}un_G(M)$. Such a measure should depend on the groupoid only up to equivalence. This is afforded by means of the concept of *groupoid cardinality*, which we are going to discuss now.

DEFINITION 4.2. A *covering map of groupoids* is a functor $F\colon \Gamma \to \Omega$ of groupoids which is surjective on objects and which satisfies the following *unique path lifting property*: For every morphism $p\colon y_1 \to y_2$ in Ω and every object x_1 in Γ for which $F(x_1)=y_1$, there exists a unique morphism $p'\colon x_1 \to x_2$ in Γ such that $F(p')=p$.

A covering map is called *n-sheeted*, with $n \in \mathbb{N}$, if the preimage of every object in Ω consists of n objects in Γ.

EXERCISE 4.3. Show that any group homomorphism $H \to G$ gives rise to a functor $*//H \to *//G$ of one-object groupoids as in Example 1.24(1).

Under which conditions is such a functor a covering map?

Let $\Phi\colon \Gamma \to \Omega$ be a functor between groupoids. Since it is in general not meaningful to ask whether objects in a category are equal, the fiber $\Phi^{-1}(y)$ of (the object function of) Φ over some $y \in \Omega$ is not invariant under equivalence of groupoids. To resolve this problem, for $y \in \Omega$ we define the *homotopy fiber* of Φ over y, to be denoted by using square brackets, i.e., as $\Phi^{-1}[y]$, as the groupoid whose objects are pairs (x, g) consisting of an object $x \in \Gamma$ and a morphism $g\colon \Phi(x) \to y$ in Ω, and for which a morphism $(x, g) \to (x', g')$ in $\Phi^{-1}[y]$ is a morphism $h\colon x \to x'$ in Γ such that $g' \circ \Phi(h) = g$.

EXERCISE 4.4. Assume that the automorphism group $\mathrm{Aut}(y)$ of an object y in a groupoid Ω is finite. Show that for any functor $\Phi\colon \Gamma \to \Omega$ of groupoids the forgetful functor $\Phi^{-1}[y] \to \Gamma_y$ to the full subgroupoid $\Gamma_y \subset \Gamma$ that consists of all $x \in \Gamma$ such that $\Phi(x) \cong y$ is an $|\mathrm{Aut}(y)|$-sheeted covering map.

EXERCISE 4.5. Let $\varphi\colon H \to G$ be a group homomorphism and let

$$\Phi\colon *//H \to *//G$$

be the functor constructed in Exercise 4.3.

Compute the homotopy fiber $\Phi^{-1}[*_G]$.

PROPOSITION 4.6. There is a unique function $|\cdot|$, called *groupoid cardinality*, that assigns to an essentially finite groupoid Γ a rational number $|\Gamma|$ that obeys the following axioms:

(1) Normalization:
$|*| = 1$, where $*$ is the groupoid with one object and with one morphism (the identity morphism id_*).
(2) Homotopy invariance:
If the groupoids Γ and Ω are equivalent, then $|\Gamma| = |\Omega|$.
(3) Additivity under disjoint union of groupoids: $|\Gamma \sqcup \Omega| = |\Gamma| + |\Omega|$.
(4) Covering property:
If there is an n-sheeted covering map $\Gamma \to \Omega$, then $|\Gamma| = n\,|\Omega|$.

PROOF. When combined with Remark 1.76, homotopy invariance and additivity imply that the groupoid cardinality is completely determined by the way it behaves on one-object groupoids $\mathcal{B}G = *//G$ with G a finite group. Denote by $\mathcal{E}G$ the action groupoid for the left regular action of G on itself. The obvious functor $\mathcal{E}G \to \mathcal{B}G$ that sends the morphism $g \xrightarrow{hg^{-1}} h$ in $\mathcal{E}G$ to the morphism $hg^{-1} \in G$ in $*//G$ is a $|G|$-sheeted covering. On the other hand, $\mathcal{E}G \to *$ is an equivalence, because the left regular action is transitive and free. The normalization axiom thus gives $|\mathcal{E}G| = 1$. The covering axiom then implies that $|\mathcal{B}G| = 1/|G|$. □

REMARK 4.7. Instead of groupoid cardinality, the term *Euler characteristic* (for groupoids) also is in use.

The groupoid cardinality will play a crucial role in our exposition of Dijkgraaf-Witten theories. If the ground field $\mathbb{K}$ has characteristic zero, then the rational numbers form a subfield of $\mathbb{K}$ and the groupoid cardinality can be identified with an element of $\mathbb{K}$.

The proof of Proposition 4.6 directly leads to a formula for the groupoid cardinality:

COROLLARY 4.8. The groupoid cardinality of an essentially finite groupoid Γ is given by the rational number

$$|\Gamma| = \sum_{[g]\in\pi_0(\Gamma)} \frac{1}{|\mathrm{Aut}(g)|},$$

where $\pi_0(\Gamma)$ is the set of isomorphism classes of Γ.

The groupoid cardinality has further properties, besides those already stated, which make it a natural measure of the size of a groupoid.

EXERCISE 4.9. Let Γ and Ω be essentially finite groupoids. Show that their product $\Gamma\times\Omega$ is essentially finite, too, and that its groupoid cardinality is $|\Gamma\times\Omega| = |\Gamma|\cdot|\Omega|$.

Hint: Verify that the function $|-\times\Omega|\,/\,|\Omega|$ satisfies the axioms in Proposition 4.6.

EXAMPLE 4.10. For S a finite set and G a finite group acting on S, the groupoid cardinality of the action groupoid $S/\!/G$ is given by

$$|S/\!/G| = |S|\,/\,|G|\,. \tag{4.11}$$

A corresponding equality does, in general, *not* hold for the set-theoretic cardinality of the set-theoretic quotient.

One idea that leads to a proof of the formula (4.11) is to replace the action groupoid $S/\!/G$ by another action groupoid for which the action is free. To this end, we consider the left regular action of G on itself and the resulting diagonal G-action on $S\times G$. This action is free, and the set of orbits is given by $(S\times G)/G\cong S$. It now follows directly from Corollary 4.8 that $|(S\times G)/\!/G| = |S|$. Since the projection $S\times G\to S$ to the first factor is a G-equivariant map, we obtain a functor $(S\times G)/\!/G\to S/\!/G$ of groupoids. It can be shown that this functor is a $|G|$-sheeted covering. By the covering property one then concludes that $|S| = |(S\times G)/\!/G| = |G|\,|S/\!/G|$.

EXERCISE 4.12. Work out a (slightly) different proof of the statement made in Example 4.10 that reduces it to the orbit-stabilizer theorem.

Find a counterexample to the statement in the case that the groupoid cardinality is replaced by the set-theoretic cardinality of the set-theoretic quotient.

REMARK 4.13. From the computation in Exercise 4.12 one may infer that large orbits make large contributions to the groupoid cardinality of an action groupoid. This allows us to think of groupoid cardinality indeed as a kind of *volume* of a groupoid. Also note that the presence of nontrivial stabilizers decreases the volume; this is a crucial feature of groupoid cardinality.

EXAMPLE 4.14. Let M be a closed oriented manifold and G a finite group. By Corollary 1.77 the groupoid $\mathcal{B}un_G(M)$ of G-bundles over M is essentially finite, so that we can use the groupoid cardinality $|\mathcal{B}un_G(M)|$ to give meaning to the path integral $\int_{\mathcal{B}un_G(M)} \mathrm{d}\Phi$ in (4.1). If M is connected, then by combining Proposition 1.73(3) and Example 4.10 one sees that the groupoid cardinality of $\mathcal{B}un_G(M)$ can be expressed in terms of the fundamental group of M as

$$|\mathcal{B}un_G(M)| = \frac{|\mathrm{Hom}(\pi_1(M), G)|}{|G|}. \tag{4.15}$$

Subsuming the observations above, we are led to define the invariant

$$Z_G(M) := |\mathcal{B}un_G(M)| \tag{4.16}$$

for any closed oriented manifold M. This invariant is multiplicative under disjoint union of manifolds, as befits an invariant for closed oriented manifolds that comes from a topological field theory.

EXAMPLE 4.17. It is instructive to compute this invariant for the one-dimensional circle S^1. The fundamental group of S^1 is $\mathbb{Z}$, so that

$$\mathrm{Hom}(\pi_1(S^1), G) \cong \mathrm{Hom}(\mathbb{Z}, G) \cong G.$$

Thus Equation (4.15) gives $|\mathcal{B}un_G(S^1)| = |G|/|G| = 1$.

REMARK 4.18. The outcome of this calculation is a particular case of the general result (see, e.g., Example 4.11(1) in [**Ca**]) that there are no nontrivial state-sum topological field theories in dimension $d=1$.

EXERCISE 4.19. Compute the invariant (4.16) for M being the three-sphere S^3 and for $M = S^2 \times S^1$.

(This computation requires the knowledge of the fundamental groups of these three-manifolds; they are $\pi_1(S^3) = \{e\}$ and $\pi_1(S^2 \times S^1) = \mathbb{Z}$.)

4.2. Toward Dijkgraaf-Witten theory

Our next goal is to use locality properties of G-bundles to extend the manifold invariant $Z_G(M)$ to a topological field theory Z_G, the so-called *Dijkgraaf-Witten theory* [**DiW**]. To achieve this, we make use of the following procedure, which is often referred to as the *linearization of* (spans of) *groupoids*; again we largely follow [**BaeHW**].

DEFINITION 4.20. An *invariant function* f on a groupoid Γ is the assignment of an element $f(x) \in \mathbb{K}$ to every object $x \in \Gamma$ in such a way that $f(x) = f(y)$ if $x \cong y$ in Γ.

We denote by $\mathcal{F}(\Gamma)$ the vector space of invariant functions on Γ.

REMARK 4.21. For any object $x \in \Gamma$ we can define an invariant function δ_x by setting $\delta_x(y) := 1$ if $x \cong y$ and $\delta_x(y) := 0$ otherwise. We may regard δ_x as a "delta function" concentrated on the isomorphism class of x. For an essentially finite groupoid Γ, the set $(\delta_x)_{[x] \in \pi_0(\Gamma)}$ forms a basis of the space $\mathcal{F}(\Gamma)$ of invariant functions.

REMARK 4.22. Obviously, an invariant function on a groupoid Γ amounts to a function on the set $\pi_0(\Gamma)$ of isomorphism classes of Γ. Furthermore, for any pair of groupoids Γ and Ω there is a distinguished isomorphism $\mathcal{F}(\Gamma) \otimes_{\mathbb{K}} \mathcal{F}(\Omega) \cong \mathcal{F}(\Gamma \times \Omega)$.

Next we observe that given a functor $\Phi\colon \Gamma \longrightarrow \Gamma'$ of groupoids, we get a pullback of invariant functions, the linear map

$$\Phi^* : \quad \mathcal{F}(\Gamma') \longrightarrow \mathcal{F}(\Gamma)\,, \qquad f' \longmapsto f' \circ \Phi\,. \tag{4.23}$$

Moreover, we have $(\Phi \circ \Phi')^* = (\Phi')^* \circ \Phi^*$ for composable functors between groupoids, and naturally isomorphic functors give rise to the same pullback map. As a consequence, the pullback map associated to an equivalence of groupoids is an isomorphism of vector spaces.

As a further ingredient of the construction of Dijkgraaf-Witten theory we need, for any functor $\Phi\colon \Gamma \longrightarrow \Gamma'$ between essentially finite groupoids, a pushforward linear map $\Phi_*\colon \mathcal{F}(\Gamma) \longrightarrow \mathcal{F}(\Gamma')$, in the direction opposite to the one of the map (4.23). If both $\mathcal{F}(\Gamma)$ and $\mathcal{F}(\Gamma')$ were endowed with the structure of a symmetric nondegenerate bilinear form $\langle - , - \rangle$, then we could define Φ_* to be the adjoint map of Φ^*, i.e., characterize Φ_* by the property

$$\langle \Phi_* f \,, f' \rangle \;=\; \langle f \,, \Phi^* f' \rangle$$

for $f \in \mathcal{F}(\Gamma)$ and $f' \in \mathcal{F}(\Gamma')$.

Now indeed, the notion of groupoid cardinality suggests a very natural choice of a symmetric nondegenerate bilinear form. To present the latter, we introduce the notion of an integral of an invariant function with respect to groupoid cardinality:

DEFINITION 4.24. Let Γ be an essentially finite groupoid and $f \in \mathcal{F}(\Gamma)$. We set

$$\int_\Gamma f \;:=\; \sum_{[x]\in\pi_0(\Gamma)} \frac{f(x)}{|\mathrm{Aut}(x)|} \;\in \mathbb{K}\,.$$

This number is called the *integral of f over Γ*.

To emphasize the terminology 'integral', we also write suggestively

$$\int_\Gamma f \equiv \int_\Gamma f(x)\,\mathrm{d}x$$

using a dummy variable x.

EXERCISE 4.25. Compute the integral over the constant function $f=1$, i.e., $f(x)=1$ for every $[x] \in \pi_0(\Gamma)$.

A few basic facts about the integral with respect to groupoid cardinality are recorded in the following exercise.

EXERCISE 4.26.

(1) Let $\Phi\colon \Gamma \longrightarrow \Omega$ be an equivalence of essentially finite groupoids, and let f be an invariant function on Ω.

   Show that the transformation formula

$$\int_\Gamma \Phi^* f \;=\; \int_\Omega f$$

   holds.

(2) Let $\Psi\colon \Gamma \longrightarrow \Omega$ be an n-sheeted covering of essentially finite groupoids. Show that for any invariant function f on Ω the formula

$$\int_\Gamma \Psi^* f \;=\; n \int_\Omega f$$

   holds.

(3) Prove the following statement, which is known as *Cavalieri's principle*:

Let $\Phi\colon \Gamma \longrightarrow \Omega$ be a functor of essentially finite groupoids. Then

$$|\Gamma| = \int_\Omega |\Phi^{-1}[y]|\,\mathrm{d}y \tag{4.27}$$

with $\Phi^{-1}[y]$ the homotopy fiber, as introduced on page 100.

LEMMA 4.28. Let $\Phi\colon \Gamma \to \Omega$ be a functor of essentially finite groupoids and f an invariant function on Γ. Then

$$\int_\Gamma f = \int_\Omega \int_{\Phi^{-1}[y]} Q_y^* f \,\mathrm{d}y\,, \tag{4.29}$$

where $Q_y\colon \Phi^{-1}[y] \to \Gamma$ is the canonical forgetful functor from the homotopy fiber over y to Γ.

PROOF. Since the integral with respect to groupoid cardinality is linear, it is sufficient to consider the case that $f = \delta_{[x]}$ for some $x \in \Gamma$, where $\delta_{[x]}$ is the function (introduced in Remark 4.21) that takes the value $1 \in \mathbb{K}$ on all objects isomorphic to x and is zero everywhere else. We can then restrict Φ to the full subgroupoid Γ_x of objects in Γ isomorphic to x and obtain

$$\int_\Gamma f = |\Gamma_x| = \int_\Omega \left|\left(\Phi|_{\Gamma_x}\right)^{-1}[y]\right| \mathrm{d}y$$

by Cavalieri's principle (4.27). But $\left(\Phi|_{\Gamma_x}\right)^{-1}[y]$ is the full subgroupoid of $\Phi^{-1}[y]$ that consists of all objects that under $Q_y\colon \Phi^{-1}[y] \longrightarrow \Gamma$ project to an object isomorphic to x. This implies the equality $\left|\left(\Phi|_{\Gamma_x}\right)^{-1}[y]\right| = \int_{\Phi^{-1}[y]} Q_y^* f$, and thus proves the claim. □

DEFINITION 4.30.

(1) Let Γ be an essentially finite groupoid. We define a symmetric nondegenerate bilinear form $\langle -\,, - \rangle$ on $\mathcal{F}(\Gamma)$ by

$$\langle f\,, f' \rangle := \int_\Gamma f(x) f'(x)\,\mathrm{d}x \tag{4.31}$$

for $f, f' \in \mathcal{F}(\Gamma)$.

(2) For a functor $\Phi\colon \Gamma \longrightarrow \Omega$ between essentially finite groupoids we denote the adjoint of the linear map Φ^* with respect to the bilinear form $\langle -\,, - \rangle$ by Φ_*.

The following result provides a more explicit expression for the adjoint Φ_*:

PROPOSITION 4.32. For any functor $\Phi\colon \Gamma \longrightarrow \Omega$ between essentially finite groupoids we have

$$(\Phi_* f)(y) = \int_{\Phi^{-1}[y]} Q_y^* f = \int_{\Phi^{-1}[y]} f(x)\,\mathrm{d}(x,h) \tag{4.33}$$

for $f \in \mathcal{F}(\Gamma)$ and $y \in \Omega$, where $Q_y\colon \Phi^{-1}[y] \longrightarrow \Gamma$ is the canonical forgetful functor from the homotopy fiber over y to Γ.

(The dummy variable (x,h) on the right-hand side of (4.33) is in $\Phi^{-1}[y]$ and thus is a pair consisting of $x \in \Gamma$ and $h\colon \Phi(x) \xrightarrow{\cong} y$.)

PROOF. For any function $f \in \mathcal{F}(\Gamma)$ we define $f'(y) := \int_{\Phi^{-1}[y]} f(x)\,\mathrm{d}(x,h)$ for $y \in \Omega$. Then $f' \in \mathcal{F}(\Omega)$, and for $g \in \mathcal{F}(\Omega)$ we get

$$\begin{aligned}\langle f', g\rangle &= \int_\Omega \Big(\int_{\Phi^{-1}[y]} f(x)\,\mathrm{d}(x,h) \Big)\, g(y)\;\mathrm{d}y \\ &= \int_\Omega \Big(\int_{\Phi^{-1}[y]} f(x)\, g(\Phi(x))\,\mathrm{d}(x,h) \Big)\;\mathrm{d}y \\ &= \int_\Omega \Big(\int_{\Phi^{-1}[y]} Q_y^*(f(\Phi^* g)) \Big)\;\mathrm{d}y \;=\; \int_\Gamma f(\Phi^* g) \;=\; \langle f\,, \Phi^* g\rangle\,.\end{aligned}$$

Here the last step uses Lemma 4.28. By the nondegeneracy of $\langle -,-\rangle$ and the definition of Φ_* as the adjoint of Φ^*, this implies that $f' = \Phi_* f$, which yields the assertion. □

EXERCISE 4.34. Show that for an equivalence $\Phi\colon \Gamma \longrightarrow \Omega$ of essentially finite groupoids, the maps Φ^* and Φ_* are inverse isomorphisms.

Hint: First verify that the pullback along a fully faithful functor preserves the scalar product.

REMARK 4.35. From the composition law for pullback maps and Definition 4.30 we deduce that

$$(\Phi' \circ \Phi)_* = \Phi'_* \circ \Phi_*$$

for any pair of composable functors between essentially finite groupoids.

REMARK 4.36. If the ground field $\mathbb{K}$ is the field $\mathbb{C}$ of complex numbers, then $\langle -\,,-\rangle$ is not a scalar product, because it is bilinear rather than sesquilinear. A scalar product is instead given by $\int_\Gamma \overline{f(x)}\, f'(x)\,\mathrm{d}x$, where the overline denotes complex conjugation.

Next recall from Definition 1.80 the notion of a (co)span. Consider a span

$$\begin{array}{ccccc} & & \Gamma & & \\ & \overset{s}{\swarrow} & & \overset{t}{\searrow} & \\ \Gamma_1 & & & & \Gamma_2 \end{array}$$

of essentially finite groupoids. By applying pullback and pushforward maps this yields the diagram

$$\begin{array}{ccccc} & & \mathcal{F}(\Gamma) & & \\ & \overset{s^*}{\nearrow} & & \overset{t_*}{\searrow} & \\ \mathcal{F}(\Gamma_1) & & & & \mathcal{F}(\Gamma_2) \end{array}$$

of vector spaces. We thus get the following linear map, called a *pull-push map*:

$$\mathcal{F}(\Gamma_1) \xrightarrow{\;s^*\;} \mathcal{F}(\Gamma) \xrightarrow{\;t_*\;} \mathcal{F}(\Gamma_2)\,.$$

Explicitly we have

$$(t_* s^* f)(\gamma_2) = \sum_{\substack{[y]\in\pi_0(\Gamma) \\ \text{such that } t(y)\cong\gamma_2}} \frac{|\mathrm{Aut}(\gamma_2)|}{|\mathrm{Aut}(y)|}\, f(s(y))$$

for $f \in \mathcal{F}(\Gamma_1)$ and $\gamma_2 \in \Gamma_2$.

With the pull-push maps at hand, we are now in a position to provide the assignments that underlie the construction of a d-dimensional Dijkgraaf-Witten topological field theory.

DEFINITION 4.37. Let G be a finite group. We refer to the following prescriptions, which assign structures in the category $\mathcal{V}ect(\mathbb{K})$ of vector spaces to structures in the cobordism category $\mathcal{C}ob_{d,d-1}$, as the *Dijkgraaf-Witten assignment*, denoted by Z_G:

(1) To a closed oriented $(d-1)$-manifold Σ, Z_G assigns the vector space of invariant functions on the groupoid of G-bundles on Σ, i.e.,

$$\Sigma \longmapsto Z_G(\Sigma) := \mathcal{F}(\mathcal{B}un_G(\Sigma))\,. \tag{4.38}$$

(2) To a morphism $M\colon \Sigma_1 \longrightarrow \Sigma_2$ in $\mathcal{C}ob_{d,d-1}$, seen as a *cospan*

$$\begin{array}{ccc} & M & \\ \nearrow & & \nwarrow \\ \Sigma_1 & & \Sigma_2 \end{array}$$

Z_G assigns a linear map

$$Z_G\big(\Sigma_1{\rightarrow}M{\leftarrow}\Sigma_2\big):\quad Z_G(\Sigma_1) \longrightarrow Z_G(\Sigma_2)$$

by the following prescription: Since bundles pull back, the inclusion maps for the incoming and outgoing boundary of M induce restriction functors, in the opposite direction, for the corresponding groupoids of G-bundles:

$$\begin{array}{ccc} & \mathcal{B}un_G(M) & \\ {}^{s}\swarrow & & \searrow^{t} \\ \mathcal{B}un_G(\Sigma_1) & & \mathcal{B}un_G(\Sigma_2) \end{array} \tag{4.39}$$

The linear map $Z_G(M) \equiv Z_G\big(\Sigma_1{\rightarrow}M{\leftarrow}\Sigma_2\big)$ is then defined as the pull-push map

$$\mathcal{F}(\mathcal{B}un_G(\Sigma_1)) \xrightarrow{\;s^*\;} \mathcal{F}(\mathcal{B}un_G(M)) \xrightarrow{\;t_*\;} \mathcal{F}(\mathcal{B}un_G(\Sigma_1))\,. \tag{4.40}$$

These assignments form the basis of a topological field theory, which will be called the (d-dimensional) *Dijkgraaf-Witten theory* for the group G, and which we will still denote by the symbol Z_G. To establish that the assignments do yield a topological field theory, we must show that they define a *functor* from $\mathcal{C}ob_{d,d-1}$ to the category of vector spaces and, furthermore, that they can be supplemented in such a way that this functor is *symmetric monoidal*. For being able to do so we still need some additional preparation. A proof that Z_G can indeed be complemented to a topological field theory will therefore have to wait until Section 4.4.

EXAMPLE 4.41. It is instructive to spell out Definition 4.37(1) for the case of one-dimensional Dijkgraaf-Witten theory. A G-bundle over a point is a G-torsor and is isomorphic to the trivial bundle. Hence the groupoid of G-bundles has a single isomorphism class of objects and its space of gauge invariant functions is one-dimensional. As a consequence, by Theorem 3.1 one-dimensional Dijkgraaf-Witten corresponds to a one-dimensional vector space. This holds for any finite group G. As a consistency check, recall from Section 3.1 that the dimension of this vector space equals the invariant $|\mathcal{B}un_G(S^1)|$ of the circle. And indeed in Example 4.17 we have seen that $|\mathcal{B}un_G(S^1)| = 1$.

REMARK 4.42. Given a morphism $\Sigma_1 \to M \leftarrow \Sigma_2$ in $\mathcal{C}ob_{d,d-1}$, we can fix a G-bundle $P \in \mathcal{B}un_G(\Sigma_1)$ and consider the invariant delta function $\delta_P \in \mathcal{F}(\mathcal{B}un_G(\Sigma_1))$ considered in Remark 4.21. In the basis given by the delta functions, $Z_G(M)$ can be expressed through its matrix elements $Z_G(M)_{P,P'}$ which are defined by the expansion

$$(Z_G(M))(\delta_P) = \sum_{[P'] \in \pi_0(\mathcal{B}un_G(\Sigma_2))} Z_G(M)_{P,P'}\, \delta_{P'}\,.$$

The numbers $Z_G(M)_{P,P'}$ can be seen as a mathematically precise incarnation of the matrix elements discussed in Section 2.1.

REMARK 4.43. It should be appreciated that, as a matter of principle, the pushforward leads to an integral. Cobordisms are cospans of manifolds. After linearization, they lead to spans of vector spaces. To turn these cospans into maps, a pushforward is used at a certain point; thereby integrals are introduced. This can be seen as the conceptual origin of "path" integrals in Dijkgraaf-Witten theories.

4.3. Two-dimensional Dijkgraaf-Witten theory

Before developing tools that allow us to prove that the assignments made in the previous section produce a topological field theory, let us familiarize ourselves with those definitions in the two-dimensional case. In that case we can describe the situation entirely in algebraic terms, thanks to the classification of two-dimensional field theories via commutative Frobenius algebras presented in Theorem 3.14. If the prescription Z_G given in Section 4.2 indeed yields a topological field theory, then there must exist a commutative Frobenius algebra associated with Z_G. We will now determine this commutative Frobenius algebra explicitly. With the help of this algebra, we will be able to prove that at least in the two-dimensional case the Dijkgraaf-Witten theory Z_G is actually a topological field theory. This is a welcome plausibility check before approaching the proof in the general d-dimensional case in Section 4.4.

We now introduce a commutative Frobenius algebra, which will afterward be shown to correspond to two-dimensional Dijkgraaf-Witten theory. Let G be a finite group G and $\mathbb{K}$ be a field that, besides being algebraically closed (which we assume throughout), has characteristic zero. We can endow the finite-dimensional vector space $\mathrm{Map}(G,\mathbb{K})$ of $\mathbb{K}$-valued functions on G with the *convolution product*, which is defined by

$$(f_1 * f_2)(g) := \sum_{\substack{g_1,g_2 \in G \\ g_1 g_2 = g}} f_1(g_1)\, f_2(g_2) \tag{4.44}$$

for $f_1, f_2 \colon G \to \mathbb{K}$ and $g \in G$.

Since G is finite, the *group delta functions* δ_g that are given, for $g \in G$, by

$$\delta_g(h) = \delta_{g,h} \qquad \text{for } h \in G\,, \tag{4.45}$$

furnish a basis $(\delta_g)_{g\in G}$ of $\mathrm{Map}(G,\mathbb{K})$. On this basis the convolution product (4.44) takes the simple form

$$\delta_g * \delta_h = \delta_{gh}\,.$$

Thus the vector space $\mathrm{Map}(G,\mathbb{K})$ endowed with the convolution product is isomorphic to the group algebra $\mathbb{K}[G]$ as introduced in Definition 1.15. In particular, the convolution product is not commutative, unless G is abelian.

That the functions δ_g form a basis implies that the center

$$A_G := \mathrm{Z}(\mathrm{Map}(G,\mathbb{K}),*)$$

of $\mathrm{Map}(G,\mathbb{K})$ consists of those functions $f \in \mathrm{Map}(G,\mathbb{K})$ which satisfy $\delta_g * f = f * \delta_g$ for every $g \in G$. This, in turn, is the case if and only if $f(gh) = f(hg)$ for all $g, h \in G$. In other words, the center A_G consists of those functions which are constant on conjugacy classes, i.e., of the *class functions* on G. One further verifies that the normalized evaluation ε of a class function at the unit element $e \in G$, i.e.

$$\varepsilon(f) := \frac{f(e)}{|G|}\,, \tag{4.46}$$

and thus $\varepsilon(\delta_g) = \delta_{g,e}/|G|$, provides a Frobenius structure on the algebra $\mathrm{Map}(G,\mathbb{K})$, with a comultiplication that is given by

$$\Delta(\delta_g) = |G| \sum_{h\in G} \delta_{gh} \otimes \delta_{h^{-1}}$$

for all $g \in G$. This restricts to a comultiplication on the center A_G and furnishes a commutative Frobenius algebra structure on A_G. (Note that this is *not* the comultiplication considered in Example 2.76 which is used to endow the group algebra with the structure of a Hopf algebra; the Hopf algebra comultiplication does not use the multiplication on the group.)

Since $\mathbb{K}$ is algebraically closed and of characteristic zero, the algebra A_G is semisimple. Schur's orthogonality relations for the characters of G-modules imply that a basis $\{e_i\}$ of commuting idempotents in A_G is given by suitable multiples of the simple characters χ_i of G:

$$e_i = \frac{\chi_i(e)}{|G|}\, \chi_i\,.$$

Thus we have

$$\varepsilon(e_i) = \frac{\chi_i(e)^2}{|G|^2}\,.$$

Now by virtue of the classification result for two-dimensional topological field theories (Theorem 3.14) we can associate to the commutative Frobenius algebra $A_G = \mathrm{Z}(\mathbb{K}[G])$ a two-dimensional topological field theory; we denote this theory by $Z_{[A_G]}$. Specifically, invoking the formula (3.17) we see that the number – also often referred to as a *partition function* – associated to a closed connected oriented surface Σ is

$$Z_{[A_G]}(\Sigma) = \sum_{\substack{\chi \text{ simple} \\ \text{character of } G}} \left(\frac{\chi(e)}{|G|}\right)^{\mathrm{Eu}(\Sigma)}, \tag{4.47}$$

with $\mathrm{Eu}(\Sigma)$ the Euler characteristic of Σ. Now note that to a simple character of G we may associate an irreducible representation V of G whose dimension is the value of the character at the neutral element e. This defines a bijection between the set of simple characters of G and the set $\mathrm{Irrep}(G)$ of isomorphisms classes of irreducible representations. Accordingly we can rewrite (4.47) equivalently as

$$Z_{[A_G]}(\Sigma) = \sum_{V\in\mathrm{Irrep}(G)} \left(\frac{\dim V}{|G|}\right)^{\mathrm{Eu}(\Sigma)} . \tag{4.48}$$

PROPOSITION 4.49. The evaluation of the two-dimensional topological field theory $Z_{[A_G]}$ on the pair of pants $S^1 \sqcup S^1 \to S^1$ and on the disk $S^1 \to \emptyset$ coincides with the assignments made for Z_G in Definition 4.37, up to natural isomorphism.

(Here we merely compare vector spaces and linear maps, because we have not yet established that Z_G can actually be complemented to a topological field theory.)

PROOF. We start by determining the vector space $Z_G(S^1)$ that the functor Z_G associates to the circle S^1. Recall that, by Example 1.74, the groupoid $\mathcal{B}un_G(S^1)$ of G-bundles over the circle is equivalent to the action groupoid $G//G$ of the action of G on itself via conjugation. (Beware that this identification depends on the choice of the base point on S^1). As a consequence we have

$$Z_G(S^1) = \big[\mathcal{B}un_G(S^1), \mathcal{V}ect(\mathbb{K})\big] \cong [G//G, \mathcal{V}ect(\mathbb{K})] . \tag{4.50}$$

and the isomorphism classes of G-bundles on S^1 are given by

$$\pi_0(\mathcal{B}un_G(S^1)) = \pi_0(G//G) = G/_{\mathrm{adj}\,G} ,$$

i.e., by the conjugacy classes of G. The space of functions on the set of conjugacy classes of a group G is by definition the space of class functions, which we have seen to coincide with A_G.

Next we compute the linear map

$$Z_G(T): \quad Z_G(S^1) \otimes Z_G(S^1) \longrightarrow Z_G(S^1)$$

that Z_G associates to the pair of pants, or trinion, T, i.e., to the sphere with three boundary components, two of them incoming and one outgoing. To this end we must evaluate $\mathcal{B}un_G(-)$ on the cospan $S^1 \sqcup S^1 \to T \leftarrow S^1$ and apply the pull-push procedure.

For doing so we need to make the identification of $\mathcal{B}un_G(S^1)$ with $G//G$ explicit, which requires in particular to fix a standard model for the circle S^1. As such we take the unit circle $\{|z|=1\} \subset \mathbb{C}$ in the complex plane, and as a base point on the unit circle we take the point $-\mathrm{i}$ on the negative imaginary axis. We need to fix a model for the pair of pants as well; we do so as follows. Remove from the unit disk in the complex plane two disks of radius $1/4$ centered at the points $\pm 1/2$, with their boundary circles regarded as incoming and the boundary circle of the unit disk regarded as outgoing. As a base point for this model of T we take the point $p_T := -\mathrm{i}/2$.

Our first task is now to map the standard model of S^1 to each of the three boundary circles of the standard model of T. On those circles we select as respective base points the points $p_0 := -\,\mathrm{i}$ and $p_\pm := \pm 1/2 - \mathrm{i}/4$, as illustrated in the following picture:

(4.51)

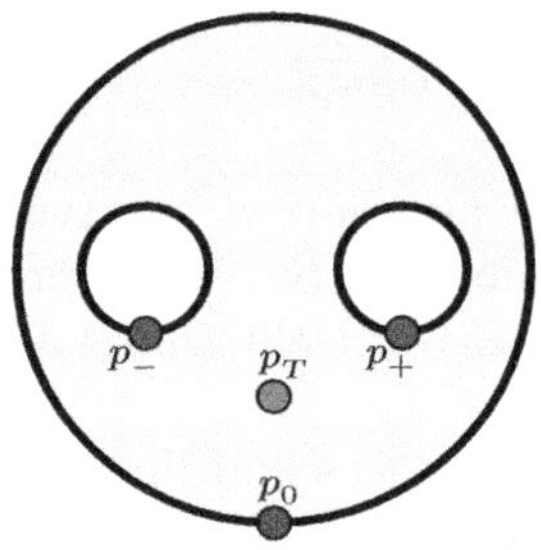

We admit only base-point preserving maps. This requirement leaves us with a contractible collection of maps, and choosing one of them does not affect holonomies. The fundamental group $\pi_1(T, p_T)$ of the pair of pants is the free group on two generators. As a representative pair of generators of $\pi_1(T, p_T)$ we choose two closed paths a and b starting at the base point p_T of T that encircle counter-clockwise either of the incoming boundary circles as well as paths (drawn as dashed lines) that connect their base points $p_\pm$ to p_T. (The analogous path encircling the outgoing boundary circle and the dashed path connecting p_0 to p_T is homotopic to the concatenation of a and b.) This is shown in the following picture:

(4.52)

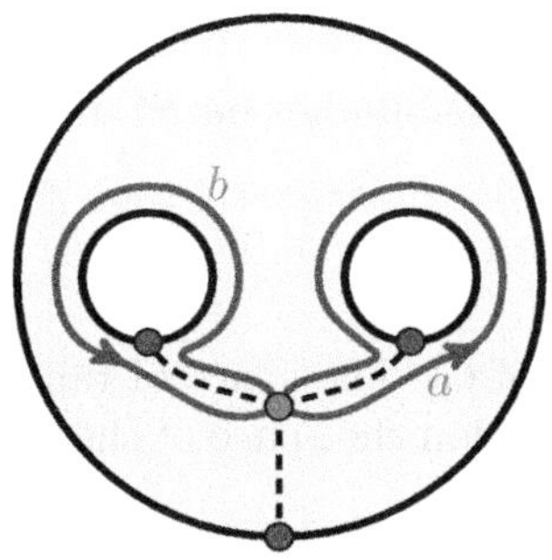

A G-bundle on T is thus determined by the holonomies $g_a, g_b \in G$ along the paths a and b. This identifies the category $\mathcal{B}un_G(T)$ with the finite action groupoid $(G \times G)//G$ in which G acts by simultaneous conjugation on the two copies of G. The dashed paths that connect the base point p_T of T to the base points p_0 and $p_\pm$ of the three boundary circles determine a homomorphism from $\pi_1(T, p_T)$ with respect to p_T to the fundamental groups of the boundary circles with respect to their base points. This allows us to describe the three relevant pullback functors $\mathcal{B}un_G(T) \longrightarrow \mathcal{B}un_G(S^1)$ as the functors that are defined on objects by $(g_a, g_b) \longmapsto g_a$ and $(g_a, g_b) \longmapsto g_b$ for the two incoming circles, and as $(g_a, g_b) \longmapsto g_a g_b$ for the outgoing circle. We conclude that the span of groupoids to which we have to apply the pull-push procedure is, up to equivalence, given by

$$
G//G \times G//G \xleftarrow{\;p\;} (G \times G)//G \xrightarrow{\;m\;} G//G. \tag{4.53}
$$

Here in the action groupoid $(G\times G)//G$ the group G acts by simultaneous conjugation on both copies of G. The functors p and m map objects as

$$(g_1,g_2) \overset{p}{\longleftarrow\!\shortmid} (g_1,g_2) \overset{m}{\longmapsto} g_1\,g_2\,;$$

on morphisms, which are gauge transformations, m is the identity, while p is the diagonal map.

Finally we note that an invariant function (that is, a function constant on orbits of the G-action) $f_1\otimes f_2\colon G\times G\longrightarrow \mathbb{K}$ on the groupoid $G//G\times G//G$ pulls back to an invariant function on $(G\times G)//G$, which is the same function, but with invariance only considered with respect to the G-action restricted to the diagonal action. The homotopy fiber over h consists of all pairs (g_1,g_2) with $g_1g_2=h$ (and hence is in particular discrete), so that by applying the formula (4.33) we see that $f_1\otimes f_2$ is pushed forward to the function f on $G//G$ for which

$$f(h) = \sum_{\substack{g_1,g_2\in G\\ g_1g_2=h}} f_1(g_1)\,f_2(g_2)$$

for $h\in G$ with invariance under the adjoint action. This is precisely the convolution product (4.44) on the algebra of class functions.

Through similar considerations one can compute the linear map that Z_G assigns to the cap, i.e., to the disk seen as cobordism $S^1\longrightarrow\emptyset$. Since the disk is contractible, there is only a single isomorphism class of G-bundles, namely (the class of) the trivial bundle. It restricts to the trivial bundle with holonomy e on its boundary. This implies that $Z_G(\mathrm{cap})([\delta_g])=\delta_{g,e}/|G|$. This reproduces the counit (4.46). □

It follows from Proposition 4.49 that Z_G is actually a two-dimensional topological field theory, whereby the comparison between $Z_{[A_G]}$ and Z_G becomes an isomorphism of topological field theories.

As a specific application of the result in Proposition 4.49, one obtains a simple proof of Mednykh's formula [**Med**]. We refer to [**Sny**] for an overview on proofs of this formula using topological field theory.

COROLLARY 4.54 (Mednykh's Formula). *For any finite group G and any connected closed surface Σ the equality*

$$\sum_{V\in\mathrm{Irrep}(G)} (\dim V)^{\mathrm{Eu}(\Sigma)} = |G|^{\mathrm{Eu}(\Sigma)-1}\,\big|\mathrm{Hom}(\pi_1(\Sigma),G)\big|$$

holds, where the summation is over the irreducible representations of G on vector spaces over a fixed algebraically closed field of characteristic zero.

PROOF. By Proposition 4.49, $Z_{[A_G]}$ and Z_G are isomorphic topological field theories. As a consequence, they yield the same numerical invariant for any connected closed surface Σ. This leads to the chain

$$\begin{aligned}\sum_{V\in\mathrm{Irrep}(G)} \Big(\frac{\dim V}{|G|}\Big)^{\mathrm{Eu}(\Sigma)} &\overset{(4.48)}{=} Z_{[A_G]}(\Sigma) \;=\; Z_G(\Sigma)\\ &\overset{(4.16)}{=} |\mathcal{B}un_G(\Sigma)| \overset{(4.15)}{=} \frac{|\mathrm{Hom}(\pi_1(\Sigma),G)|}{|G|}\end{aligned}$$

of equalities. □

EXERCISE 4.55. Evaluate Mednykh's Formula in the case that Σ is the torus.

4.4. Dijkgraaf-Witten theory as a topological field theory

In this section we develop further tools that will enable us to establish that the assignments prescribed in Definition 4.37 can be complemented in such a way that they actually furnish a topological field theory in the sense of Definition 2.100. This result will be stated as Theorem 4.66. Some readers may want to skip the technical details that enter the proof, but in any case they should, in addition to the main theorem, also take notice of the gluing property for principal bundles as described in Proposition 4.64. In our exposition we follow again essentially the arguments in [**FrQ**, **BaeHW**, **Mor3**].

The geometric input for the construction of Dijkgraaf-Witten theory are the groupoids of G-bundles for a finite group G on varying manifolds. In order to establish that Dijkgraaf-Witten theory is actually a functor, we need to prove that it translates the gluing of cobordisms to the composition of linear maps. To this end, we will have to understand the groupoid of G-bundles for a manifold that has been obtained by gluing. A convenient tool for approaching such problems in a conceptual manner is the language of stacks [**Hei**]. Here we will instead take a more elementary approach, which uses the following notion:

DEFINITION 4.56. Consider a cospan

$$\Gamma \xrightarrow{\Phi} \Omega \xleftarrow{\Psi} \Lambda$$

of groupoids. The *homotopy pullback* $\Gamma \times_\Omega \Lambda$ of such a cospan is the groupoid whose objects are triples (x, y, η) consisting of objects $x \in \Gamma$ and $y \in \Lambda$ and an isomorphism

$$\eta : \quad \Phi(x) \xrightarrow{\cong} \Psi(y)\,. \tag{4.57}$$

A morphism $(x, y, \eta) \longrightarrow (x', y', \eta')$ is a pair consisting of a morphism $f\colon x \longrightarrow x'$ in the groupoid Γ and a morphism $g\colon y \longrightarrow y'$ in Λ such that $\eta' \circ \Phi(f) = \Psi(g) \circ \eta$. The homotopy pullback $\Gamma \times_\Omega \Lambda$ comes with two projection functors $\pi_\Gamma\colon \Gamma \times_\Omega \Lambda \longrightarrow \Gamma$ and $\pi_\Lambda\colon \Gamma \times_\Omega \Lambda \longrightarrow \Lambda$ such that by the assignment $(x, y, \eta) \longmapsto \eta$ we get a natural isomorphism $\Phi \circ \pi_\Gamma \longrightarrow \Psi \circ \pi_\Lambda$. Drawing this natural isomorphism as a diagram of the form introduced in (1.51) we get the square

$$\begin{array}{ccc} \Gamma \times_\Omega \Lambda & \xrightarrow{\pi_\Gamma} & \Gamma \\ {\scriptstyle \pi_\Lambda}\downarrow & \overset{\eta}{\swarrow\!\!\!\swarrow} & \downarrow{\scriptstyle \Phi} \\ \Lambda & \xrightarrow{\Psi} & \Omega . \end{array} \tag{4.58}$$

We express this situation by saying that the square (4.58) of groupoids *commutes up to* the natural isomorphism η, and also refer to the double arrow for η as a *filler* of the square.

REMARK 4.59. The homotopy pullback should be compared to the pullback that appears, e.g., in Exercise 1.45. The latter involves an equality, whereas in the situation at hand we have instead an isomorphism. This is in accordance with the general principle formulated in Remark 1.12.

REMARK 4.60. Taking Λ in Definition 4.56 to be the category $*//\mathrm{id}_*$ with a single object and a single morphism (namely the identity of that object), the functor $\Psi\colon *//\mathrm{id}_* \longrightarrow \Omega$ is just an object $y \in \Omega$, and the corresponding homotopy pullback is just the homotopy fiber $\Phi^{-1}[y]$ of the functor Φ over y.

EXERCISE 4.61. Consider the homotopy pullback (4.58) from Definition 4.56, with the same notation as used there.

Prove that for $y \in \Lambda$ there is a canonical equivalence

$$\pi_\Lambda^{-1}[y] \xrightarrow{\simeq} \Phi^{-1}[\Psi(y)]$$

between homotopy fibers.

The following crucial statement relates the homotopy pullback to the pullback and pushforward of invariant functions:

PROPOSITION 4.62 (Beck-Chevalley property). For every cospan $\Gamma \xrightarrow{\Phi} \Omega \xleftarrow{\Psi} \Lambda$ of essentially finite groupoids, the square

$$\begin{array}{ccc} \mathcal{F}(\Gamma) & \xrightarrow{\Phi_*} & \mathcal{F}(\Omega) \\ {\scriptstyle \pi_\Gamma^*}\big\downarrow & & \big\downarrow{\scriptstyle \Psi^*} \\ \mathcal{F}(\Gamma \times_\Omega \Lambda) & \xrightarrow{\pi_{\Lambda *}} & \mathcal{F}(\Lambda) \end{array} \tag{4.63}$$

of vector spaces of invariant functions commutes.

PROOF. For any $f \in \mathcal{F}(\Gamma)$ and $y \in \Lambda$, by Proposition 4.32 we have

$$(\Psi^* \Phi_* f)(y) = \int_{\Phi^{-1}[\Psi(y)]} Q_{\Psi(y)}^* f \,,$$

where $Q_{\Psi(y)} \colon \Phi^{-1}[\Psi(y)] \to \Gamma$ is the forgetful functor. By Exercise 4.61 there is a canonical equivalence $\pi_\Lambda^{-1}[y] \xrightarrow{\simeq} \Phi^{-1}[\Psi(y)]$. Applying the transformation formula from Exercise 4.26(1) to that equivalence gives us

$$(\Psi^* \Phi_* f)(y) = \int_{\pi_\Lambda^{-1}[y]} V_y^* \pi_\Gamma^* f = (\pi_{\Lambda *} \pi_\Gamma^* f)(y) \,,$$

with the forgetful functor $V_y \colon \pi^{-1}[y] \to \Gamma \times_\Omega \Lambda$. □

We will also need a gluing property for principal fiber bundles. The proof of this property requires some (albeit elementary) covering theory and is thus somewhat beyond the scope of these lecture notes. Nonetheless we present at least an overview of the ideas entering it. (Readers who are willing to accept the statement of Proposition 4.64 may skip these details and jump directly to the proposition.) Let $X\colon Y_0 \to Y_1$ and $X'\colon Y_1 \to Y_2$ be composable morphisms in the d-dimensional cobordism category. The commutative square

$$\begin{array}{ccc} Y_1 & \longrightarrow & X \\ \downarrow & & \downarrow \\ X' & \longrightarrow & X' \circ X \end{array}$$

that is formed by the inclusions is a pushout in the category of topological spaces. Invoking the fact that assigning to a principal bundle its transport operator gives an equivalence of categories (see Proposition 1.73), it suffices to show that the functor $[\Pi_1(-), G\text{-}\mathcal{T}\!or]$ from the opposite category of topological spaces to groupoids sends this pushout to a homotopy pullback. In a first step, the Seifert-van-Kampen theorem for fundamental groupoids (see Theorem 6.7.2 of [**Bro**]) asserts that the diagram

$$\begin{array}{ccc} \Pi_1(Y_1) & \longrightarrow & \Pi_1(X) \\ \downarrow & & \downarrow \\ \Pi_1(X') & \longrightarrow & \Pi_1(X' \circ X) \end{array}$$

is a pushout in groupoids. Since all of the functors involved are injective on objects, the diagram is also a homotopy pushout. (The latter is the homotopically correct version of a pushout; it is defined dually to the homotopy pullback that is explicitly given in Definition 4.56.) Now the functor $[-, G\text{-}\mathcal{T}\!or]$ maps this homotopy pushout to a homotopy pullback. Altogether we thus arrive at the following statement:

PROPOSITION 4.64 (Gluing property for principal fiber bundles). *Let $X\colon Y_0 \to Y_1$ and $X'\colon Y_1 \to Y_2$ be composable morphisms in the d-dimensional cobordism category, and let G be a group. Then the restriction functor*

$$R\colon \quad \mathcal{B}un_G(X' \circ X) \longrightarrow \mathcal{B}un_G(X) \times_{\mathcal{B}un_G(Y_1)} \mathcal{B}un_G(X')$$

is an equivalence of groupoids.

REMARK 4.65. Here it is important to treat individual bundles as objects in a groupoid, instead of working with isomorphism classes of bundles. Indeed, since an interval I is contractible, any G-bundle over I is isomorphic to the trivial bundle, so that there is only a single isomorphism class of G-bundles. On the other hand, for nontrivial G, there are nontrivial G-bundles on the circle S^1. To obtain these bundles when constructing S^1 by identifying the endpoints of I, we must account for the automorphisms of the objects, since they enter crucially when calculating the homotopy pullback after applying the functor $[-, G\text{-}\mathcal{T}\!or]$. Such a strategy is impossible at the level of isomorphism classes. In short, working with a groupoid ensures the correct behavior of gluing bundles and thus ensures locality.

We can now finally state:

THEOREM 4.66. *For any finite group G, the assignments Z_G given for objects in (4.38) and for morphisms in (4.40) can be complemented so as to define a d-dimensional topological field theory.*

The so defined topological field theory is called the (d-dimensional) *Dijkgraaf-Witten theory* for the group G.

PROOF. As a first step, we show that Z_G is a functor, and afterward that it is symmetric monoidal.

To prove that Z_G preserves identities, we consider for an object Y in the d-dimensional cobordism category $\mathcal{C}ob_{d,d-1}$ the diagram

$$\begin{array}{ccccc} & & \mathcal{B}un_G(Y\times I) & & \\ & \swarrow^{s} & \downarrow{\scriptstyle s} & \searrow^{t} & \\ \mathcal{B}un_G(Y) & & & & \mathcal{B}un_G(Y), \\ & \nwarrow_{\mathrm{id}} & & \nearrow_{\mathrm{id}} & \\ & & \mathcal{B}un_G(Y) & & \end{array}$$

where s and t are the restriction of G-bundles to the incoming and outgoing boundary, respectively, of the cylinder $Y \times I$. This diagram commutes up to natural isomorphism. Since naturally isomorphic functors give rise to the same pullback map and hence also pushforward map, it follows that

$$Z_G(Y\times I) = t_* \circ s^* = s_* \circ s^*.$$

By Exercise 4.34, this linear map is indeed the identity of $Z_G(Y)$.

Next consider composable morphisms $X\colon Y_0 \longrightarrow Y_1$ and $X'\colon Y_1 \longrightarrow Y_2$ in $\mathcal{C}ob_{d,d-1}$. Then we obtain the following diagram of essentially finite groupoids:

$$\begin{array}{ccccccc} & & & \mathcal{B}un_G(X'\circ X) & & & \\ & & \swarrow^{r} & \downarrow{\scriptstyle R} & \searrow^{r'} & & \\ & & & \mathcal{B}un_G(X)\times_{\mathcal{B}un_G(Y_1)}\mathcal{B}un_G(X') & & & \\ & & \swarrow^{p} & & \searrow^{p'} & & \\ & \mathcal{B}un_G(X) & & & & \mathcal{B}un_G(X') & \\ \swarrow^{s} & & \searrow^{t} & & \swarrow^{s'} & & \searrow^{t'} \\ \mathcal{B}un_G(Y_0) & & & \mathcal{B}un_G(Y_1) & & & \mathcal{B}un_G(Y_2) \end{array} \tag{4.67}$$

Here the inner square is a homotopy pullback square and commutes up to natural isomorphism. All functors in the diagram (4.67) except for the projections p and p' in the homotopy pullback square arise from restriction of bundles. As a consequence the outer square and the two triangles commute strictly. The functor R from Proposition 4.64 is an equivalence. We can thus compute

$$\begin{aligned}
Z_G(X' \circ X) &= (t' \circ r')_* \circ (s \circ r)^* \\
&= (t' \circ p' \circ R)_* \circ (s \circ p \circ R)^* && \text{(by commutativity of the triangles)} \\
&= t'_* \circ p'_* \circ R_* \circ R^* \circ p^* \circ s^* && \text{(by Eq. after (4.23) and Remark 4.35)} \\
&= t'_* \circ p'_* \circ p^* \circ s^* && \text{(by Exercise 4.34)} \\
&= t'_* \circ s'^* \circ t_* \circ s^* && \text{(by Proposition 4.62)} \\
&= Z_G(X') \circ Z_G(X) \,.
\end{aligned}$$

Finally, we note that the composition

$$\begin{aligned}
Z_G(Y_1 \sqcup Y_2) &= \mathcal{F}(\mathcal{B}un_G(Y_1 \sqcup Y_2)) \\
&\cong \mathcal{F}(\mathcal{B}un_G(Y_1) \times \mathcal{B}un_G(Y_2)) && \text{(by Exercise 1.64)} \\
&\cong \mathcal{F}(\mathcal{B}un_G(Y_1)) \otimes_{\mathbb{K}} \mathcal{F}(\mathcal{B}un_G(Y_2)) && \text{(by Remark 4.22)} \\
&= Z_G(Y_1) \otimes Z_G(Y_2)
\end{aligned}$$

of isomorphisms, for any pair of objects Y_1 and Y_2 in $\mathcal{C}ob_{d,d-1}$, endow Z_G with a monoidal structure, which is also symmetric. □

CHAPTER 5

Extended topological field theories

The value of a d-dimensional topological field theory on a d-dimensional manifold may be computed by cutting X along submanifolds of codimension 1. It is clearly desirable to be able to iterate this procedure by cutting further along manifolds of codimension 2. A topological field theory admitting such an iteration enjoys a higher degree of locality. The formal implementation of this idea turns out to be a rather challenging task: We need to replace the d-dimensional cobordism category – that is, a two-layered structure built from d- and $(d-1)$-dimensional manifolds – by a three-layered structure featuring in addition $(d-2)$-dimensional manifolds.

The resulting three-layered categorical structure – called a *bicategory* – goes beyond the realm of categories. A considerable part of this chapter is devoted to introducing this concept, with the cobordism bicategory as a guiding example in mind. Moreover, in parallel with extending cobordisms to codimension 2, our standard algebraic target category, i.e., the category of vector spaces, must undergo a similar extension process. Once this has been achieved, we will be able to define the notion of an *extended* topological field theory. This procedure can be further extended with the help of higher categories which have still more layers of structure.

5.1. Bicategories

To introduce the notion of a bicategory, first recall that a category with a single object $*$ is specified by the monoid $\mathrm{End}(*)$. A *bi*category with a single object $*$ can be specified by its endomorphisms $\mathcal{E}nd(*)$ as well, but together with the composition these will form a monoidal category instead of just a monoid. As a consequence, the composition of endomorphisms of $*$ will no longer be associative "on the nose", but rather involve associators that obey constraints which generalize the pentagon axiom. We will use the abbreviated notation "1" for the category $*//\mathrm{id}_*$ with one object and one morphism (namely the identity of that object). The definition of a bicategory then takes the following form [**Lei**, **Scho**]:

DEFINITION 5.1. A *bicategory* $\mathcal{B}$ consists of the following data, subject to the axioms given below:

- A class $\mathrm{Obj}(\mathcal{B})$. Its elements are referred to as *objects* or *0-cells*.
- For every pair (A, B) of 0-cells a category $\mathcal{H}om(A, B)$.

  The objects of $\mathcal{H}om(A, B)$ are referred to as *1-cells*, or as *1-morphisms*, and the morphisms of $\mathcal{H}om(A, B)$ are called *2-cells*, or *2-morphisms*. The composition of 2-cells in the $\mathcal{H}om$ categories is referred to as *vertical composition*. By definition it is strictly associative; we denote it by the symbol $\circ$.

- For every triple $A, B, C \in \mathrm{Obj}(\mathcal{B})$ a *composition functor*

$$\begin{aligned} c_{A,B,C}: \quad \mathcal{H}om(B,C) \times \mathcal{H}om(A,B) &\longrightarrow \mathcal{H}om(A,C) \\ (g,f) &\longmapsto g \circ f \\ (\beta,\alpha) &\longmapsto \beta \star \alpha\,, \end{aligned}$$

  and for every object $A \in \mathrm{Obj}(\mathcal{B})$ an *identity functor*

$$\mathrm{I}_A: \quad 1 = *//\mathrm{id}_* \longrightarrow \mathcal{H}om(A,A)\,.$$

  The composition of 2-cells by means of this composition functor, which we denote by the symbol $\star$, is called the *horizontal composition.*
- Natural isomorphisms a, r, l of functors expressing associativity and unitality: For every quadruple $A, B, C, D \in \mathrm{Obj}(\mathcal{B})$ a natural isomorphism $a_{A,B,C,D}$ up to which the horizontal composition is associative; this is indicated by inserting $a_{A,B,C,D}$ as a filler (displayed as a double arrow, analogously as in (4.58)) in the diagram

$$\begin{array}{ccccc} & & \mathcal{H}om(C,D) \times \mathcal{H}om(B,C) \times \mathcal{H}om(A,B) & & \\ & {\scriptstyle c_{B,C,D} \times \mathrm{id}} \swarrow & & \searrow {\scriptstyle \mathrm{id} \times c_{A,B,C}} & \\ \mathcal{H}om(B,D) \times \mathcal{H}om(A,B) & & \xRightarrow{\;a_{A,B,C,D}\;} & & \mathcal{H}om(C,D) \times \mathcal{H}om(A,C). \\ & {\scriptstyle c_{A,B,D}} \searrow & & \swarrow {\scriptstyle c_{A,C,D}} & \\ & & \mathcal{H}om(A,D) & & \end{array}$$

  Further, for every pair $A, B \in \mathrm{Obj}(\mathcal{B})$ natural isomorphisms $r_{A,B}$ and $l_{A,B}$ which again provide fillers for the diagrams

$$\begin{array}{ccc} \mathcal{H}om(A,B) \times 1 & \xrightarrow{\;\simeq\;} & \\ {\scriptstyle \mathrm{id} \times \mathrm{I}_A} \downarrow & \overset{r_{A,B}}{\Nearrow} & \searrow \\ \mathcal{H}om(A,B) \times \mathcal{H}om(A,A) & \xrightarrow[c_{A,A,B}]{} & \mathcal{H}om(A,B) \end{array}$$

  and

$$\begin{array}{ccc} 1 \times \mathcal{H}om(A,B) & \xrightarrow{\;\simeq\;} & \\ {\scriptstyle \mathrm{I}_B \times \mathrm{id}} \downarrow & \overset{l_{A,B}}{\Nearrow} & \searrow \\ \mathcal{H}om(B,B) \times \mathcal{H}om(A,B) & \xrightarrow[c_{A,B,B}]{} & \mathcal{H}om(A,B). \end{array}$$

  Thus in particular for all composable 1-cells f, g and h there are 2-cells

$$\begin{aligned} a_{h,g,f}: &\qquad (h \circ g) \circ f \xrightarrow{\cong} h \circ (g \circ f)\,, \\ r_f: &\qquad f \circ \mathrm{id}_A \xrightarrow{\cong} f\,, \\ l_f: &\qquad \mathrm{id}_B \circ f \xrightarrow{\cong} f\,. \end{aligned} \tag{5.2}$$

These data are subject to the axioms that the following diagrams commute:

- A pentagon diagram

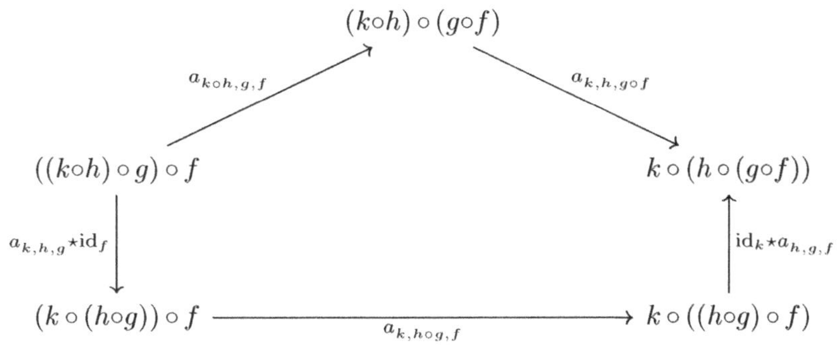

  for every composable quadruple f, g, h, k of 1-cells.

- Triangle diagrams:

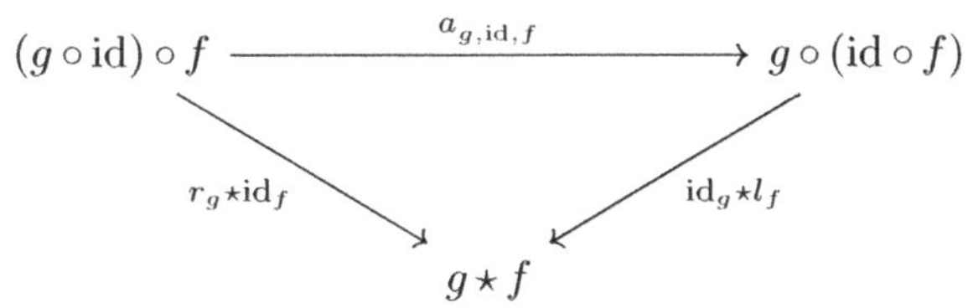

REMARK 5.3. While for 1-morphisms there is a single composition operation, for which a standard notation is '$\circ$', like for morphisms in a category, 2-morphisms can be composed in two different ways – vertically and horizontally, so that two different symbols are needed. Our choice of '$\circ$' and '$\star$' for these is one possible convention; note that we use the same symbols '$\circ$' and '$\star$' also for the vertical and horizontal composition of natural transformations. Several other conventions are in use in the literature as well.

A 2-cell α between 1-cells $f, g \in \mathcal{H}om(A, B)$ in a bicategory will be drawn as a *pasting diagram*

(5.4)

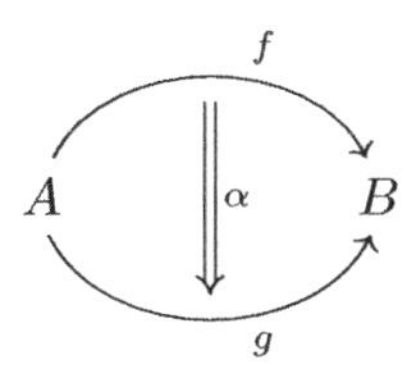

i.e., 0-cells are depicted as points, 1-cells as lines and 2-cells as surfaces.

In the discussion 2.1 of principles of field theories we have seen that the structure of a category is essential for keeping locality in gauge theories. In a similar vein, bicategories appear in generalizations of gauge theories in which one deals with "gauge transformations between gauge transformations". A geometric setting in which this is realized are *bundle gerbes*; correspondingly, the bundle gerbes over a given manifold form a bicategory rather than a category [**Wal**].

EXAMPLES 5.5.

(1) There is a bicategory $\mathcal{C}at$ having small categories as 0-cells, functors as 1-cells, and natural transformations as 2-cells.

Indeed, for two given categories $\mathcal{C}$ and $\mathcal{D}$ the functors from $\mathcal{C}$ to $\mathcal{D}$ form a category $\mathrm{Hom}(\mathcal{C}, \mathcal{D})$. The morphisms of this category are natural transformations. The composition of 1-morphisms of the bicategory of small categories is given by the the composition of functors – and thus composition happens to be strictly associative and strictly unital – and the horizontal composition of natural transformations.

Also, the pasting diagram (5.4) specializes to the diagram (1.51) for a natural transformation. Similarly, the vertical and horizontal composition of 2-cells in any bicategory can be expressed in terms of pasting diagrams analogous to the diagrams (1.52) and (1.53) for the vertical and horizontal composition of natural transformations.

(2) In particular, the collection of all essentially small groupoids forms a bicategory.

(3) Let $\mathbb{K}$ be a field. The bicategory $\mathcal{A}lg(\mathbb{K})$ has $\mathbb{K}$-algebras as objects, and for any pair A_1, A_2 of algebras, $\mathcal{H}om(A_1, A_2)$ is the category of A_1-A_2-bimodules. Composition of 1-morphisms is given by the tensor product ${}_{A_1}M_{A_2} \otimes_{A_2} {}_{A_2}M'_{A_3}$ of bimodules; the associators are induced by the associators of vector spaces and are thus not trivial. (In particular, the endomorphism category $\mathcal{H}om(A, A)$ is the monoidal category of A-bimodules, which has been described in Example 2.15(7).)

(4) Associated with any monoidal category $\mathcal{C}$ there comes a bicategory $\mathrm{B}\mathcal{C}$ that has a single object $*$ and has the monoidal category as its category of 1-endomorphisms, $\mathcal{H}om(*, *) = \mathcal{C}$. This bicategory $\mathrm{B}\mathcal{C}$ is called the *delooping* of $\mathcal{C}$.

Conversely, any bicategory with a single object is described by a monoidal category.

EXERCISE 5.6.

(1) Verify that $\mathcal{A}lg(\mathbb{K})$ as defined in Example 5.5(3) does satisfy the axioms of a bicategory.

(2) Likewise, verify that for any monoidal category $\mathcal{C}$, the delooping $\mathrm{B}\mathcal{C}$ defined in Example 5.5(4) satisfies all requirements of a bicategory.

REMARK 5.7. For bicategories there is a strictification result which extends the one (see Remark 2.46) for monoidal categories: every bicategory is biequivalent to a 2-category, i.e., to a bicategory for which the coherence 2-cells (5.2) are identities; for details, see, e.g., § 1.4 of [**GorPS**]. For *tri*categories, which have one additional layer of structure, certain strictification statements can still be made: every tricategory is equivalent to a so-called Gray-category for which, in short, everything except for the interchange law is strict. On the other hand, strict 3-categories are genuinely more restrictive than generic tricategories; see, e.g., [**Gur**] and § 5.6 of [**GorPS**]. For instance, a braided monoidal category can be regarded as a tricategory with a single object and a single 1-morphism. Such a tricategory can be strictified if and only if the braiding is in fact symmetric.

EXERCISE 5.8. For any given category $\mathcal{C}$, consider the following two monoids: the monoid $(M,\circ) := (\mathrm{End}(\mathrm{id}_{\mathcal{C}}),\circ)$ of endotransformations of the identity functor $\mathrm{id}_{\mathcal{C}}$ of $\mathcal{C}$ equipped with vertical composition, and the monoid $(M,\star) := (\mathrm{End}(\mathrm{id}_{\mathcal{C}}),\star)$ of endotransformations of $\mathrm{id}_{\mathcal{C}}$ with horizontal composition.

Show that these two monoids are equal. Conclude that the monoid M is commutative.

Hint: Show that horizontal and vertical composition can be interchanged in the sense that

$$(a\circ b)\star(c\circ d) = (a\star c)\circ(b\star d) \qquad \text{for all } a,b,c,d\in M\,.$$

(This is known as the *Eckmann-Hilton identity*.)

GRAPHICAL DESCRIPTION 5.9. The string diagram for a morphism

$$f\in\mathrm{Hom}(X,Y)$$

in a category, as introduced in Graphical Description 1.16, can be seen as the Poincaré dual of the diagram $X \xrightarrow{f} Y$. Similarly one obtains a string calculus for bicategories by considering the Poincaré dual of pasting diagrams such as (5.4), whereby 0-cells are depicted as two-dimensional regions, 1-cells as lines and 2-cells as points, or rather, analogously as morphisms in the case of categories, as small coupons. Also, conveniently the labeling of 0-cells is indicated by a shading. Thus, e.g., a 1-morphism $f\colon A\to B$ is depicted as

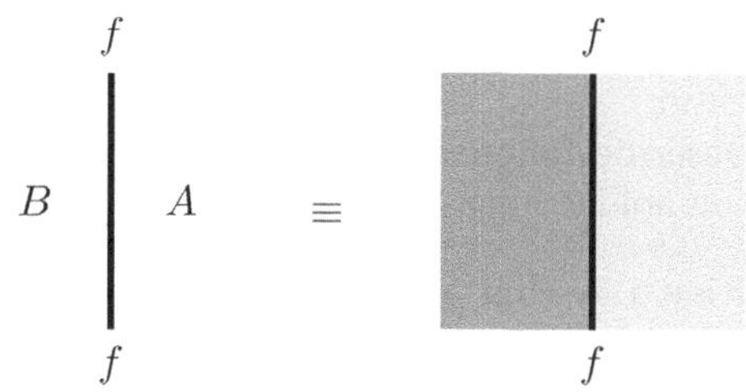

while a 2-morphism $\alpha\colon h\circ g\to f$, with pasting diagram

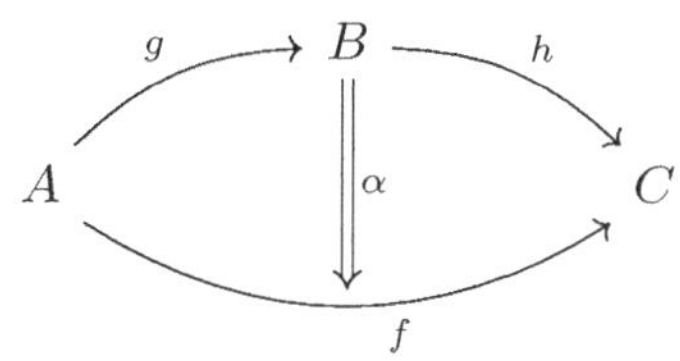

looks like

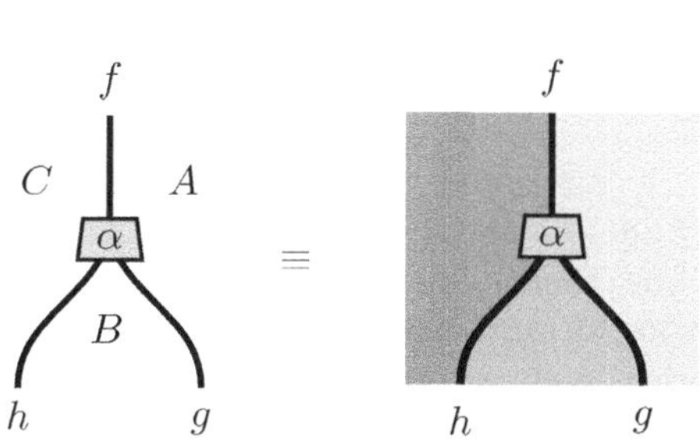

More information about the bicategorical string calculus can, e.g., be found in [**HiM**] and [**Hum**].

REMARK 5.10. For a bicategory with only a single object, no information is lost if one suppresses the object label, respectively the corresponding shading, in string diagrams altogether. Thereby they are reduced to string diagrams for a monoidal category. This fits well with the observation in Example 5.5(4) that a monoidal category is secretly a one-object bicategory. In short, the graphical string calculus for monoidal categories can be seen as a special case of the one for bicategories.

Functors between bicategories have an an additional layer of structure as well. We only sketch their definition; for the details see [**Lei**]: A *functor* $F\colon \mathcal{B}\to\mathcal{B}'$ between bicategories $\mathcal{B}$ and $\mathcal{B}'$ consists of

- the assignment of an object $F(A)$ in $\mathcal{B}'$ to every object A in $\mathcal{B}$;
- a functor $F_{A,B}\colon \mathcal{H}om_{\mathcal{B}}(A,B)\to\mathcal{H}om_{\mathcal{B}'}(F(A),F(B))$ for every pair (A,B) of objects in $\mathcal{B}$;
- invertible 2-cells $\phi_A\colon \mathrm{id}_{F(A)}\to F(\mathrm{id}_A)$ for every object A, as well as

$$\phi_{g,f}: \quad F(g)\circ F(f)\longrightarrow F(g\circ f)$$

  for every pair g and f of composable 1-morphisms.

The 2-cells are subject to appropriate coherence conditions; see, e.g., Chapter 2 of [**Gur**].

REMARK 5.11. What we call a *functor* between bicategories is in the literature sometimes instead referred to as a *pseudofunctor*, or also, in the strict case, as a *2-functor*.

EXERCISE 5.12. Describe functors $\mathrm{B}\mathcal{C}\to\mathrm{B}\mathcal{D}$ between one-object bicategories and compare them to monoidal functors $\mathcal{C}\to\mathcal{D}$.

Recall from Definition 1.47 that a natural transformation α between two functors $F,G\colon \mathcal{C}\to\mathcal{D}$ that have the same domain and same codomain assigns to any object $X\in\mathcal{C}$ a morphism $\alpha_X\colon F(X)\to G(X)$. In the same spirit, given two functors $F,G\colon \mathcal{A}\to\mathcal{B}$ between the same bicategories, we introduce the notion of a *pseudonatural transformation*: A pseudonatural transformation assigns to each object $a\in\mathcal{A}$ a 1-morphism $\alpha_a\colon F(a)\to G(a)$ and to each 1-morphism $f\colon a\to a'$ in $\mathcal{A}$ an invertible 2-morphism α_f in $\mathcal{B}$ that is a filler in the diagram

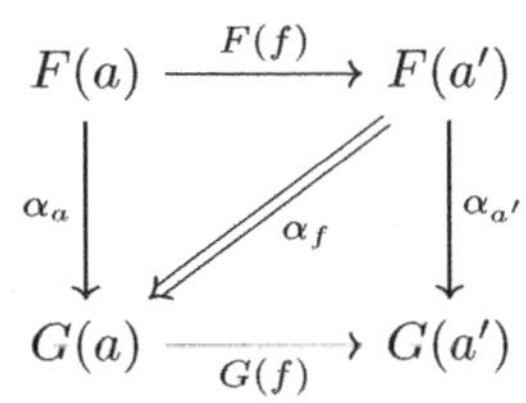

for every 1-morphism $f\colon a\to a'$ in $\mathcal{A}$. Pseudonatural transformations have to satisfy two coherence axioms; their explicit form is, e.g., given on page 5 of [**Lei**].

A pseudonatural transformation between two *monoidal* functors between the same monoidal categories, seen as one-object bicategories (see Exercise 5.12) involves an object of the monoidal category and a family of morphisms. Thus in this case, not every pseudonatural transformation comes from a monoidal natural transformation.

Finally, in the case of bicategories there is a third layer of structure: Given two pseudonatural transformations $\alpha, \beta\colon F \longrightarrow G$ with $F, G\colon \mathcal{A} \longrightarrow \mathcal{B}$, a *modification* $m\colon \alpha \longrightarrow \beta$ assigns to each object $a \in \mathcal{A}$ a 2-morphism $m_a\colon \alpha_a \longrightarrow \beta_a$ in $\mathcal{B}$, subject to a further relation, which is displayed on page 6 of [**Lei**].

Schematically, the assignments given by functors, natural transformations and modifications can be summarized as follows:

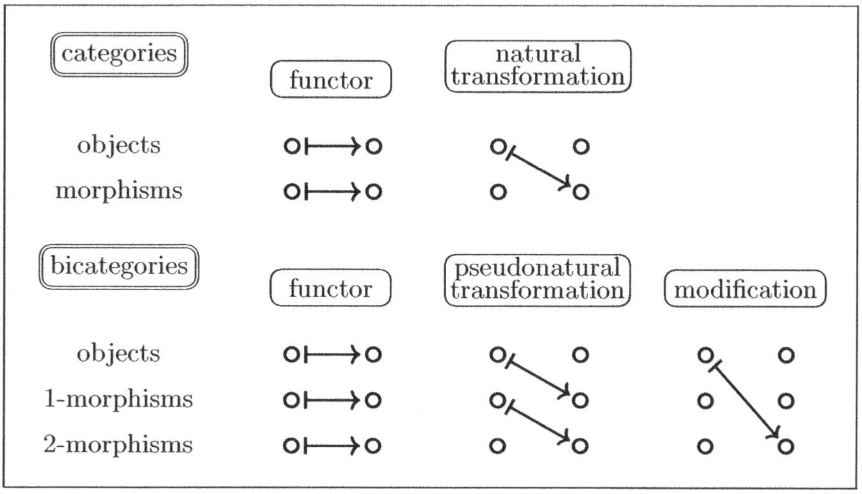

REMARK 5.13. There is also a notion of a *monoidal bicategory*. Here we give just the main idea, referring for the details to Chapter 3 of [**Scho**]: A monoidal structure on a bicategory $\mathcal{C}$ is a functor $\otimes\colon \mathcal{C} \times \mathcal{C} \longrightarrow \mathcal{C}$ of bicategories (as sketched above) that, as a binary operation on $\mathcal{C}$, is associative and unital up to specified 1-morphisms (which are again called *associators* and *unitors*) which, in turn, are invertible up to invertible 2-morphisms. The associators satisfy a pentagon axiom analogously as for monoidal categories, but the pentagon commutes only up to an invertible 2-morphism, a so-called *pentagonator*. In other words, the conditions appearing in the definition of a monoidal category get promoted to structure.
Similarly there is the notion of a *monoidal functor* $F\colon \mathcal{C} \longrightarrow \mathcal{D}$ between monoidal bicategories, i.e., a functor between monoidal bicategories that is compatible with the monoidal structures. In particular, such a functor comes equipped with 1-morphisms $F(c) \otimes F(c') \longrightarrow F(c \otimes c')$ for $c, c' \in \mathcal{C}$ and a 1-morphism $\mathbf{1}_{\mathcal{D}} \longrightarrow F(\mathbf{1}_{\mathcal{C}})$ for the monoidal units $\mathbf{1}_{\mathcal{C}}$ and $\mathbf{1}_{\mathcal{D}}$ of $\mathcal{C}$ and $\mathcal{D}$. These 1-morphisms are invertible up to invertible 2-isomorphisms.
One can also define the notion of a braiding on a monoidal bicategory and the notion of a braided monoidal functor. A braided monoidal bicategory can have a symmetric structure, but unlike for ordinary categories, the symmetry of the braiding is now structure and not just a property. Finally, one can define the notion of a symmetric monoidal functor between symmetric monoidal bicategories.

5.2. Toward Dijkgraaf-Witten theory as an extended TFT

We now consider three-dimensional Dijkgraaf-Witten theories; our goal is to incorporate locality at a deeper level and to extract interesting algebraic structure.

We start with the following observation. Iterating the argument that led us to consider three-dimensional manifolds with two-dimensional boundaries, we continue to chop the two-manifolds into smaller pieces as well, and thereby introduce two-manifolds with boundary into the picture. (In principle one can even further iterate this extension procedure, compare Remark 5.19(5) below.) Thus we drop the requirement on the two-manifold Σ to be closed and instead allow Σ to be a compact oriented two-manifold with one-dimensional boundary $S := \partial\Sigma$.

Next we investigate heuristically what type of quantity we must associate to a one-manifold S with a specified field configuration $\phi_1 \in \mathcal{B}un_G(S)$. For any surface Σ with boundary S it is natural to consider the space of all field configurations $\mathcal{B}un_G(\Sigma, \phi_1)$ on Σ that restrict to the given field configuration ϕ_1 on the boundary $S \cong \partial\Sigma$, similarly as we did in the prescription (2.5).

Once we have fixed boundary values, we can again linearize the situation. For a two-manifold Σ with boundary $\partial\Sigma = S$, we consider for any field configuration ϕ_1 on S the vector space $\mathcal{H}_{\Sigma,\phi_1}$ that is freely generated by the isomorphism classes of field configurations on Σ with given boundary value ϕ_1.

Given a two-manifold with boundary S, we hereby obtain a map $\phi_1 \mapsto \mathcal{H}_{\Sigma,\phi_1}$ from field configurations to vector spaces. Thus for any two-manifold Σ with boundary S we get a complex vector bundle over the space of all fields on the boundary.

In the case of a finite group G, we prefer to regard such a vector bundle as an object in the category $[\mathcal{B}un_G(S), \mathcal{V}ect(\mathbb{K})]$ of functors from $\mathcal{B}un_G(S)$ to $\mathcal{V}ect(\mathbb{K})$. Accordingly we should associate to the one-dimensional manifold S precisely this linear category of vector bundles:

$$Z_G(S) := [\mathcal{B}un_G(S), \mathcal{V}ect(\mathbb{K})] \,.$$

This is not simply a collection of vector spaces parametrized by the set $\pi_0(\mathcal{B}un_G(S))$. Rather, if a bundle has automorphisms, then these act on the corresponding vector space. An explicit description of the category $Z_G(S^1)$ obtained for the circle S^1 will be provided in Proposition 5.33.

These observations motivate us to extend our discussion to include codimension-2 objects in the definition of a topological field theory as well. We thus arrive at a three-layered structure, involving symmetric monoidal bicategories.

DEFINITION 5.14.

(1) A *2-vector space* (over a field $\mathbb{K}$) is a $\mathbb{K}$-linear abelian finitely semisimple category.
(2) The *bicategory* 2-$\mathcal{V}ect(\mathbb{K})$ *of* 2-*vector spaces* has 2-vector spaces as objects, $\mathbb{K}$-linear functors as 1-morphisms, and natural transformations as 2-morphisms.

The bicategory 2-$\mathcal{V}ect(\mathbb{K})$ can be endowed with the structure of a symmetric monoidal bicategory (see Remark 5.13). The symmetric monoidal structure is given by the *Deligne tensor product* $\boxtimes$ of finite abelian categories. For a detailed description of the Deligne tensor product we refer to Section 5 of [**Del**] and Chapter 1.11 of [**EtGNO**]. For our purposes the following aspects are sufficient:

- The tensor product $\otimes_{\mathbb{K}}$ of vector spaces can be characterized in terms of a universal property relating bilinear maps $V \times W \longrightarrow Z$ to linear maps $V \otimes_{\mathbb{K}} W \longrightarrow Z$.

  In the same vein, the Deligne tensor product $\mathcal{C} \boxtimes \mathcal{D}$ of $\mathbb{K}$-linear finite abelian categories can be characterized by relating right exact $\mathbb{K}$-linear functors $\mathcal{C} \times \mathcal{D} \longrightarrow \mathcal{X}$ to right exact $\mathbb{K}$-linear functors $\mathcal{C} \boxtimes \mathcal{D} \longrightarrow \mathcal{X}$.

- If, for $i = 1, 2$, A_i-mod is a category of finite-dimensional modules over a finite-dimensional algebra A_i in $\mathcal{V}ect(\mathbb{K})$, then one has

$$A_1\text{-mod} \boxtimes A_2\text{-mod} \cong (A_1 \otimes_{\mathbb{K}} A_2)\text{-mod}\,.$$

(Here it is used that $\mathbb{K}$ is algebraically closed and hence perfect.)
- The unit for the tensor product $\boxtimes$ is the category $\mathbb{K}$-mod $= \mathcal{V}ect$.

In the spirit of Definition 2.100, we want to define an extended topological field theory as a symmetric monoidal functor from a cobordism bicategory to the algebraic bicategory 2-$\mathcal{V}ect(\mathbb{K})$. When defining the relevant cobordism bicategory one must be careful, because one may not identify any longer diffeomorphic two-manifolds. We only give an outline of the full definition:

Definition 5.15. $\mathcal{C}ob_{3,2,1}$ is the following symmetric monoidal bicategory:

- Objects are compact closed oriented one-manifolds S.
- 1-morphisms are two-dimensional compact oriented collared cobordisms $S \times I \hookrightarrow \Sigma \hookleftarrow S' \times I$.
- 2-morphisms are generated by diffeomorphisms of cobordisms that fix the collar and by three-dimensional collared oriented cobordisms with corners M.
- Both horizontal and vertical composition are by gluing along collars.
- The monoidal structure is given by disjoint union, with the empty set $\emptyset$ as the monoidal unit.

Remarks 5.16. In the complete definition, more care is in particular needed to properly account for the following aspects:

(1) Cobordisms with corners are considered up to diffeomorphisms that preserve the orientation and boundary.
(For an example of a cobordism with corners see the picture (5.26) below.)

(2) For manifolds with corners, the specification of the gluing along collars, and thus of the compositions in $\mathcal{C}ob_{3,2,1}$, involves quite a few more details than in the case of $\mathcal{C}ob_{3,2}$, i.e., manifolds with boundary. For instance, one must work with the notion of a *germ of neighborhoods* of a manifold. For a detailed exposition see Section 4.3 of [**Mor2**] and Sections 3.1 and 3.2 of [**Scho**].

(3) To account for the relevance of collars one should, in the treatment of the bicategory $\mathcal{C}ob_{d,d-1,d-2}$ for any d, not consider embeddings $S_{d-2} \longrightarrow \Sigma_{d-1}$, but instead what we call *collared embeddings*, i.e., diffeomorphisms $f\colon S_{d-2} \times [0, \epsilon) \longrightarrow \Sigma_{d-1}$ such that $\mathrm{Im}\,(f) \cap \partial\Sigma_{d-1} = \mathrm{Im}\,(f(-,0))$, and furthermore identify collared embeddings that coincide on some connected neighborhood of $S_{d-2} \times \{0\}$. (Thus one actually deals with a germ of neighborhoods: the value of the number $\epsilon > 0$ does not matter.) Using the symbol '$\twoheadrightarrow$' to denote a collared embedding, for a d-manifold M_d with corners in an extended cobordism category we deal with

embeddings that fit into a diagram of the form

(5.17)

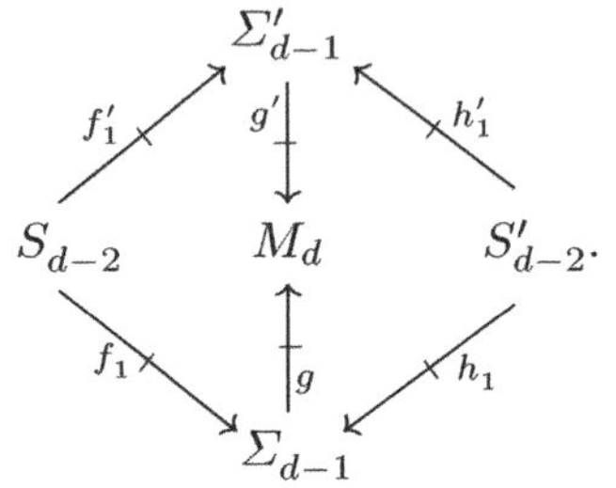

The manifold M_d contains embedded parametrized tubes

$$\tau\colon\ S_{d-2}\times[0,1]\longrightarrow M_d \qquad \text{and} \qquad \tau'\colon\ S'_{d-2}\times[0,1]\longrightarrow M_d$$

obeying $\tau(s,t)=g'\circ f'_1(s)$ and $\tau(s,1-t)=g\circ f_1(s)$ for $s\in S_{d-2}$ and $t\in[0,\epsilon)$, and similarly for τ'. We call this prescription which makes use of collared embeddings and parametrized tubes the *principle of globularity.*

We are now ready to present the definition of a (3-2-1-)extended topological field theory:

DEFINITION 5.18. A *(once-)extended three-dimensional topological field theory*, also called 3-2-1-*dimensional topological field theory*, is a symmetric monoidal functor

$$Z:\quad \mathcal{C}ob_{3,2,1}\longrightarrow 2\text{-}\mathcal{V}ect(\mathbb{K})\,.$$

To justify the qualification *extended* in this definition, we make the following observations:

REMARKS 5.19.

(1) As a monoidal functor, Z must send the monoidal unit $\emptyset$ in the bicategory $\mathcal{C}ob_{3,2,1}$ to the monoidal unit $\mathcal{V}ect(\mathbb{K})$ in $2\text{-}\mathcal{V}ect(\mathbb{K})$. The functor Z restricts to a functor $Z|_\emptyset$ from the endomorphisms of $\emptyset$ in $\mathcal{C}ob_{3,2,1}$ to the endomorphisms of $\mathcal{V}ect(\mathbb{K})$ in $2\text{-}\mathcal{V}ect(\mathbb{K})$.

(2) It follows directly from the definition that

$$\mathrm{End}_{\mathcal{C}ob_{3,2,1}}(\emptyset)\;\simeq\;\mathcal{C}ob_{3,2}$$

as symmetric monoidal categories. Using the fact that the morphisms in $2\text{-}\mathcal{V}ect(\mathbb{K})$ are additive (which follows from $\mathbb{K}$-linearity of functors in the definition of 2-vector spaces), it is also easy to see that there is an equivalence

$$\mathrm{End}_{2\text{-}\mathcal{V}ect(\mathbb{K})}\big(\mathcal{V}ect(\mathbb{K})\big)\;\simeq\;\mathcal{V}ect(\mathbb{K})$$

of symmetric monoidal categories. This equivalence maps a linear endofunctor of the category of vector spaces to its value on the ground field $\mathbb{K}$.

(3) Hence we can conclude: If Z is an extended three-dimensional topological field theory, then $Z|_\emptyset$ is a three-dimensional topological field theory in the sense of Definition 2.100.

(4) This raises the question of whether a given (nonextended) three-dimensional topological field theory can be extended. In general, there is no reason for this to be possible. In contrast, for Dijkgraaf-Witten theories such an extension can

be constructed with the help of principal bundles and their linearization. Thus there exists an extended three-dimensional topological field theory Z_G which assigns the category

$$[\mathcal{B}un_G(S), \mathcal{V}ect(\mathbb{K})]$$

to the one-dimensional closed oriented manifold S and whose restriction $Z_G|_\emptyset$ is (isomorphic to) the Dijkgraaf-Witten theory described in Theorem 4.66.

(5) One can iterate the procedure of extension and introduce the notion of a *fully extended topological field theory* which in particular also assigns quantities to points. We refer to [**Lur**] for a discussion of fully extended topological field theories and their relation to the *cobordism hypothesis*. In [**FrHLT**] an argument is sketched that a Dijkgraaf-Witten theory can be turned into a fully extended topological field theory. A discussion of fully extended topological field theories is beyond the scope of these lectures.

5.3. Construction via 2-linearization

In this subsection we present the construction of the 3-2-1-dimensional Dijkgraaf-Witten theory for a finite group G. The construction, which is described in [**FrQ, Mor1, Mor3**], is a bicategorical generalization of the strategy already applied in the nonextended case: One first assigns a bundle groupoid to every manifold that appears in the cobordism category and then linearizes. The bicategorical version of the two-step procedure turns out to be more involved than in the nonextended situation. It therefore makes sense to separate the two steps and to introduce the symmetric monoidal bicategory of spans of groupoids; for details we refer to [**Mor1**], and for additional information to [**Hau**].

DEFINITION 5.20. The bicategory $\mathcal{S}pan$ of *spans of groupoids* is the following symmetric monoidal bicategory:

- Objects Γ are essentially finite groupoids.
- 1-morphisms $\Gamma_1 \to \Gamma_2$ are spans of essentially finite groupoids, i.e., diagrams

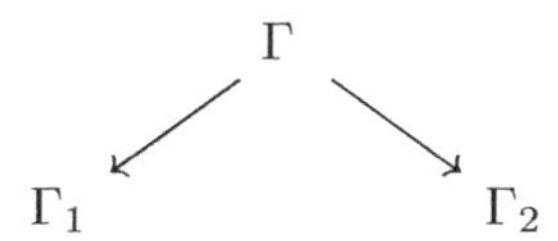

of essentially finite groupoids.
- 2-morphisms are equivalence classes of spans of spans

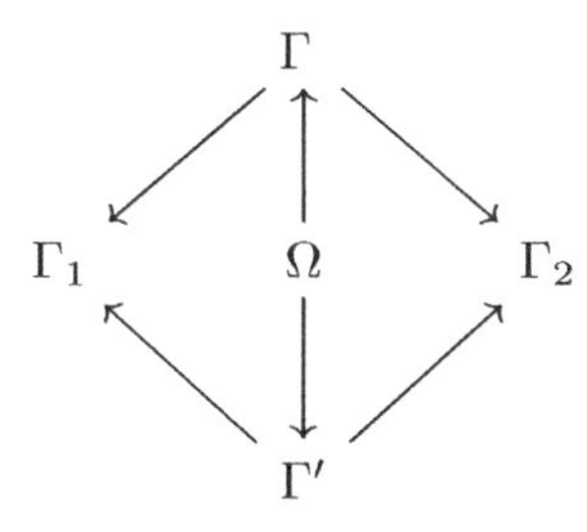

- Horizontal composition is given by forming homotopy pullbacks of groupoids. Spelled out in more detail, the composition of two 1-morphisms $\Gamma_1 \leftarrow \Gamma \to \Gamma_2$ and $\Gamma_2 \leftarrow \Gamma' \to \Gamma_3$ is the homotopy pullback $\Gamma \times_{\Gamma_2} \Gamma'$ equipped with natural maps $\Gamma \times_{\Gamma_2} \Gamma' \longrightarrow \Gamma \longrightarrow \Gamma_1$ and $\Gamma \times_{\Gamma_2} \Gamma' \longrightarrow \Gamma' \longrightarrow \Gamma_3$. The composition of 2-morphisms is defined similarly.

- The monoidal structure is given by the Cartesian product of groupoids.

The idea is now to obtain a symmetric monoidal functor with codomain $\mathcal{S}pan$ by taking the bundle groupoids of all the manifolds appearing in $\mathcal{C}ob_{3,2,1}$: We can assign to a one-dimensional manifold S the groupoid $\mathcal{B}un_G(S)$ of G-bundles over S, to a two-manifold $\Sigma\colon S \longrightarrow S'$ with parametrized boundary the span

$$\begin{array}{ccc} & \mathcal{B}un_G(\Sigma) & \\ \swarrow & & \searrow \\ \mathcal{B}un_G(S) & & \mathcal{B}un_G(S') \end{array}$$

and to a three-dimensional cobordism M with corners between two-manifolds $\Sigma, \Sigma'\colon\ S \longrightarrow S'$ with parametrized boundaries and parametrized tubes (adhering to the principle of globularity discussed in Remark 5.16(3)) the span of spans of groupoids given by

$$\begin{array}{ccccc} & & \mathcal{B}un_G(\Sigma) & & \\ & \swarrow & \uparrow & \searrow & \\ \mathcal{B}un_G(S) & & \mathcal{B}un_G(M) & & \mathcal{B}un_G(S'). \\ & \nwarrow & \downarrow & \nearrow & \\ & & \mathcal{B}un_G(\Sigma') & & \end{array} \tag{5.21}$$

All functors needed here come from restriction of G-bundles. The diagram (5.21) does, in general, not commute on the nose; for achieving commutativity, natural transformations are needed as fillers. These result from the fact that the two maps $S \longrightarrow M$ are only homotopic by globularity, so that by the mechanism described in Example 1.56 we obtain natural transformations between pullback functors for the left quadrangle in (5.21). For the right quadrangle the same conclusion follows from the homotopy between the two maps $S' \longrightarrow M$.

The functor $\mathcal{B}un_G(-)$ sends, by Proposition 4.64, the composition of cobordisms to homotopy pullbacks, and disjoint unions to Cartesian products. We thus obtain

PROPOSITION 5.22. The assignments just described yield a symmetric monoidal functor

$$\mathcal{B}un_G : \quad \mathcal{C}ob_{3,2,1} \longrightarrow \mathcal{S}pan\,.$$

The second step in the construction of extended Dijkgraaf-Witten theory is the 2-*linearization*: We have already explained the idea to associate to an essentially finite groupoid Γ its category of vector bundles $\mathcal{V}ect(\Gamma)$ on Γ, i.e., the functor category $\big[\Gamma, \mathcal{V}ect(\mathbb{K})\big]$. We next need to explain what 2-linearization assigns to spans of groupoids.

The assignment

$$\Gamma \mapsto \mathcal{V}ect(\Gamma) := [\Gamma, \mathcal{V}ect(\mathbb{K})]$$

is a contravariant functor from the bicategory of essentially finite groupoids to the bicategory of 2-vector spaces. Functors $\Gamma \xrightarrow{f} \Gamma'$ between groupoids are sent to pullback functors:

$$\begin{aligned} [\Gamma', \mathcal{V}ect] &\longrightarrow [\Gamma, \mathcal{V}ect], \\ \varphi &\longmapsto \varphi \circ f. \end{aligned}$$

LEMMA 5.23. [**Mor1**, 4.2.1] Let $f\colon \Gamma \to \Gamma'$ be a functor between essentially finite groupoids. Then the pullback functor $f^*\colon \mathcal{V}ect(\Gamma') \to \mathcal{V}ect(\Gamma)$ admits a *two-sided* adjoint

$$f_*: \quad \mathcal{V}ect(\Gamma) \longrightarrow \mathcal{V}ect(\Gamma')$$

The functor f_* is called the *pushforward.*

We use this pushforward to associate to a span

$$\Gamma \xleftarrow{p_0} \Lambda \xrightarrow{p_1} \Gamma'$$

of essentially finite groupoids the *'pull-push' functor*

$$(p_1)_* \circ (p_0)^* : \quad \mathcal{V}ect_{\mathbb{K}}(\Gamma) \longrightarrow \mathcal{V}ect_{\mathbb{K}}(\Gamma').$$

A similar construction given in detail in [**Mor1**] associates to spans of spans a natural transformation.

Altogether we have:

PROPOSITION 5.24. The functor $\Gamma \mapsto \mathcal{V}ect_{\mathbb{K}}(\Gamma)$ extends to a symmetric monoidal functor

$$\mathcal{V}_{\mathbb{K}}: \quad \mathcal{S}pan \longrightarrow 2\text{-}\mathcal{V}ect(\mathbb{K})$$

from the bicategory of spans of groupoids to the bicategory of 2-vector spaces.

The functor $\mathcal{V}_{\mathbb{K}}$ is called *2-linearization*.

We are now in a position to define Dijkgraaf-Witten theory as a three-dimensional extended topological field theory by concatenating the symmetric monoidal functors from Propositions 5.22 and 5.24:

DEFINITION 5.25. Dijkgraaf-Witten theory as an extended three-dimensional topological field theory is the composition

$$\mathcal{Z}_G := \mathcal{V}_{\mathbb{K}} \circ \mathcal{B}un_G : \quad \mathcal{C}ob_{3,2,1} \longrightarrow 2\text{-}\mathcal{V}ect(\mathbb{K})$$

of the symmetric monoidal functors $\mathcal{B}un_G$ and $\mathcal{V}_{\mathbb{K}}$.

It remains to be shown that $\mathcal{Z}_G|_\emptyset$ coincides with the three-dimensional Dijkgraaf-Witten theory from Theorem 4.66. This can be achieved by explicit computation, as done in [**Mor3**, Section 5.2].

5.4. Evaluation on the circle

By realizing cobordisms in terms of generators and relations (compare Section 3.2 for the case of $\mathcal{C}ob_{2,1}$) one can obtain the values of a topological field theory Z on arbitrary manifolds from its values on the generators, i.e., on a small number of particularly simple manifolds, which include, e.g., the circle S^1 as a one-dimensional generator and the pair of pants as a two-dimensional one. Evaluating Z on such simple manifolds is therefore of particular interest.

Specifically, a 3-2-1-dimensional extended topological field theory Z involves the following assignments to these generators:

Z assigns to the circle S^1 a $\mathbb{K}$-linear abelian finitely semisimple category

$$\mathcal{C}_Z := Z(S^1) .$$

To the pair of pants, Z associates a functor

$$\otimes : \quad \mathcal{C}_Z \boxtimes \mathcal{C}_Z \longrightarrow \mathcal{C}_Z .$$

This turns out to provide a tensor product, in the sense of Definition 2.13, on the category $\mathcal{C}_Z$.

Next consider a three-dimensional manifold of the following form:

(5.26)

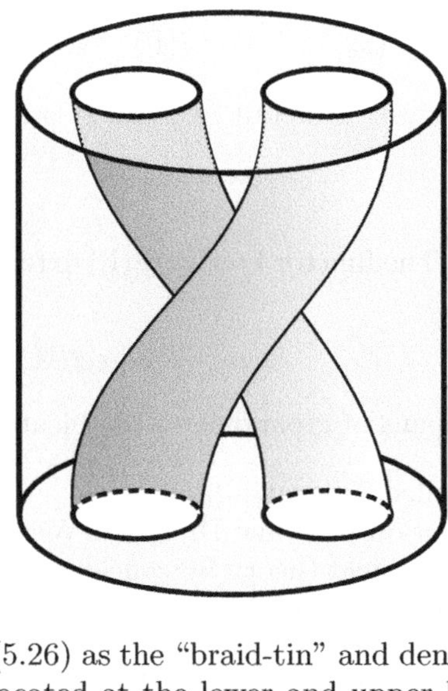

We refer the manifold (5.26) as the "braid-tin" and denote it by X_{br}. (For each of the two pairs of pants located at the lower and upper lid of the braid-tin, the outer boundary component of the two-holed disk is the outgoing circle, while the two inner boundary components are the incoming ones.) The three-manifold X_{br} (5.26) furnishes a 2-morphism between two copies of a pair of pants. To specify this 2-morphism we describe X_{br} as being obtained by evolving the standard model T for the pair of pants, as displayed in the picture (4.51), along an interval $[0,1]$, with $T\times\{0\}$ and $T\times\{1\}$ the bottom and top lid of the tin, respectively, and prescribe for each of the boundary circles S^1_i of T a homotopy from $S^1_i\times\{0\}\subset T\times\{0\}$ to $S^1_i\times\{1\}\subset T\times\{1\}$.

In the case of the outgoing boundary circle we do so by defining the map $S^1_{\rm out} \times [0,1] \to T \times [0,1]$ as $({\rm e}^{{\rm i}\,\alpha},t) \mapsto ({\rm e}^{{\rm i}\,\alpha},t)$ for $\alpha \in [0,2\pi)$ and $t \in [0,1]$, while for the two incoming boundary circles we set

$$(\pm\frac{1}{2}+\frac{1}{4}\,{\rm e}^{{\rm i}\,\alpha},t) \longmapsto (\pm\frac{1}{2}\,{\rm e}^{{\rm i}\,\pi t}+\frac{1}{4}\,{\rm e}^{{\rm i}\,\alpha},t)$$

for $\alpha \in [0,2\pi)$ and $t \in [0,1]$. To the so defined cobordism $X_{\rm br}$, the topological field theory Z assigns a natural transformation

$$\otimes \implies \otimes^{\rm opp}.$$

As we will explain in the proof of Proposition 5.35, this natural transformation constitutes a braiding on $\mathcal{C}_Z$.

There is yet another piece of structure that we can extract from the topological field theory. To see this, consider a two-manifold with boundary that is an annulus. This admits an automorphism that comes from a full rotation of one of its boundary circles against the other. Evaluation of the topological field theory on this invertible 2-morphism yields a natural isomorphism

$$\theta: \quad {\rm id}_{\mathcal{C}_Z} \xrightarrow{\cong} {\rm id}_{\mathcal{C}_Z} \tag{5.27}$$

from the identity functor on $\mathcal{C}_Z$ to itself.

The categorical interpretation of this isomorphism θ is provided via the following concept:

DEFINITION 5.28.

(1) A *balanced braided category* is a braided category $\mathcal{C}$ equipped with a natural isomorphism $\theta\colon {\rm id}_\mathcal{C} \xrightarrow{\cong} {\rm id}_\mathcal{C}$ of the identity functor that satisfies $\theta_\mathbf{1} = {\rm id}_\mathbf{1}$ and

$$\theta_{X\otimes Y} = c_{Y,X} \circ c_{X,Y} \circ (\theta_X \otimes \theta_Y)$$

for every pair of objects $X, Y \in \mathcal{C}$.

Such a natural isomorphism is called a *balancing*.

(2) A *ribbon category* is a rigid balanced braided category $\mathcal{C}$ such that

$$\theta_{X^\vee} = (\theta_X)^\vee$$

for every $X \in \mathcal{C}$. In this case the balancing is also called a *ribbon structure* or a *ribbon twist*.

As suggested by the choice of notation, the natural isomorphism θ (5.27) can be shown to be a balancing. We can then summarize:

PROPOSITION 5.29. For an extended three-dimensional topological field theory Z, the category $\mathcal{C}_Z = Z(S^1)$ is naturally endowed with the structure of a balanced braided monoidal category.

The pictorial description (5.26) of the braiding on the category $\mathcal{C}_Z = Z(S^1)$ indicates in particular that one should carefully distinguish between the braiding and its inverse (or, in more colloquial wording, between over-braiding and under-braiding). And indeed, the braided category that is obtained by evaluating the extended Dijkgraaf-Witten theory on the circle is not symmetric: as will be shown in Proposition 5.46, in that case the category $\mathcal{C}_Z$ is equivalent to the category of finite-dimensional modules over the Drinfeld double $\mathcal{D}(G)$ of the group G, and in Example 5.56 we will see that the braiding on that category is nondegenerate in the sense of Definition 5.53 and thus in particular not symmetric.

REMARK 5.30. The preceding considerations make the topological origin of the tensor product, the braiding and the balancing plausible. A formal proof that this way one indeed obtains a balanced braided structure on $\mathcal{C}_Z = Z(S^1)$ can be achieved by realizing that the oriented genus-zero surfaces with multiple inputs and one output form what is called the *framed* E_2*-operad*. This endows the category $\mathcal{C}_Z = Z(S^1)$ with the structure of an algebra over the framed E_2-operad. As explained in [**Wah, SaW**] this amounts precisely to a balanced braided structure on $\mathcal{C}_Z = Z(S^1)$. An in-depth discussion of operads and their algebras is, however, beyond the scope of this text.

REMARK 5.31. In Proposition 5.29 it is understood that all pertinent structure lives in the symmetric monoidal bicategory of 2-vector spaces. That is, the underlying category is a 2-vector space, the tensor product consists of linear functors, and the braiding and balancing are natural isomorphisms between linear functors. All of this follows in fact from the very definition, because the extended topological field theory is assumed to take values in 2-vector spaces. However, in other circumstances one might want to consider different target categories, and then the issue of where the structure maps live becomes important. For stronger emphasis it would thus make sense to expand the term (*balanced*) *braided monoidal category* in the previous statements to (*balanced*) *braided monoidal category in* 2*-vector spaces.*

5.5. A balanced braided category from bundles

According to Proposition 5.29, the extended Dijkgraaf-Witten theory for a finite group G (Definition 5.25) produces, by evaluation on the circle, a balanced braided monoidal category $\mathcal{C}_{Z_G} = Z_G(S^1)$; we denote this category by $\mathcal{C}_{\mathbb{K}}(G)$. We know from (4.50) that after choosing a base point we can write

$$\mathcal{C}_{\mathbb{K}}(G) = \big[\mathcal{B}un_G(S^1), \mathcal{V}ect(\mathbb{K})\big] \cong [G/\!/G, \mathcal{V}ect(\mathbb{K})]\,, \tag{5.32}$$

where in the action groupoid $G/\!/G$ the group G acts on itself by the adjoint action. We spell out this functor category explicitly:

PROPOSITION 5.33. For the extended Dijkgraaf-Witten theory, the category $\mathcal{C}_{\mathbb{K}}(G)$ associated to the circle S^1 is, after the choice of a base point, equivalent to the category whose objects are G-graded vector spaces $V = \bigoplus_{g\in G} V_g$ together with a G-action on V such that

$$g\,.\,V_h \subseteq V_{ghg^{-1}} \tag{5.34}$$

for all $g, h \in G$, and whose morphisms are homogeneous linear maps that intertwine the G-action.

PROOF. A functor $F\colon G/\!/G \rightarrow \mathcal{V}ect(\mathbb{K})$ has to be specified on objects and on morphisms. Defining F on objects amounts to assigning to each group element $g \in G$ a vector space $V_g := F(g)$. We assemble the collection of these vector spaces to a G-graded vector space $V := \bigoplus_{g\in G} V_g$. Further, any $g \in G$ amounts to morphisms $h \rightarrow ghg^{-1}$ in the action groupoid $G/\!/G$. The functor F sends such a morphism to a linear isomorphism $\phi_h(g)\colon V_h \longrightarrow V_{ghg^{-1}}$. We assemble these linear maps for $h \in G$ to an endomorphism $\phi(g)$ of V; by construction, $\Phi(g)\colon V \longrightarrow V$ covers the adjoint action of G on itself. Finally, given two functors $F, F'\colon G/\!/G \longrightarrow \mathcal{V}ect(\mathbb{K})$ with corresponding G-graded vector spaces V and V' and actions ϕ and ϕ', respectively, according to the commuting diagram (1.48) a natural transformation $F \longrightarrow F'$ amounts precisely to a homogeneous map $V \longrightarrow V'$ that intertwines the respective actions. □

As a next step we determine the tensor product on the category $\mathcal{C}_{\mathbb{K}}(G)$. We work with the "small" realization

$$\mathcal{C}_{\mathbb{K}}(G) \cong [G/\!/G, \mathcal{V}ect(\mathbb{K})]$$

given in (5.32). Recall that this identification depends on the choice of a base point on the circle S^1. Accordingly we must specify a standard model for the circle and fix a base point. The two-manifold that yields the tensor product is a pair of pants T, and we need to specify a standard model and a base point for T as well. We have actually already made such choices in the proof of Proposition 4.49. Applying the 2-linearization $\mathcal{V}_{\mathbb{K}}$ to the results obtained there we obtain the linear functor that underlies the tensor product in $\mathcal{C}_{\mathbb{K}}(G)$. We are now in a position to give

PROPOSITION 5.35.

(1) The tensor product of objects V and W in $\mathcal{C}_{\mathbb{K}}(G)$ is the G-graded vector space whose component at $g \in G$ is

$$(V \otimes W)_g = \bigoplus_{\substack{s,t \in G \\ st=g}} V_s \otimes W_t$$

together with the diagonal G-action

$$g.(v, w) = (g.v, g.w)$$

for all $g \in G$, $v \in V$ and $w \in W$. The associators are the obvious ones induced by the tensor product in $\mathcal{V}ect(\mathbb{K})$.

(2) The balancing $V \longrightarrow V$ is given by

$$v \longmapsto g.v$$

for $v \in V_g$.

(3) The braiding $V \otimes W \longrightarrow W \otimes V$ is given by

$$v \otimes w \longmapsto g.w \otimes v$$

for $v \in V_g$ and $w \in W$.

Proof. We prove the statements (1) and (2) for the tensor product and the balancing and provide pertinent information needed for deriving the statement (3) for the braiding, leaving further details of proving the latter as an exercise.

(1) Upon linearization, the span (4.53) of groupoids yields the cospan

$$\begin{array}{ccc} & [(G\times G)/\!/G, \mathcal{V}ect(\mathbb{K})] & \\ {}^{p^*}\nearrow & & \nwarrow^{m^*} \\ [G/\!/G \times G/\!/G, \mathcal{V}ect(\mathbb{K})] & & [G/\!/G, \mathcal{V}ect(\mathbb{K})] \end{array}$$

of linear categories. Here on the right we need an arrow in the reverse direction, meaning concretely that we must use an adjoint of m^*, which by Lemma 5.23 is a two-sided adjoint, and which we denote by F.

The functor F sends a G-bigraded vector space $W = \bigoplus_{\gamma_1,\gamma_2\in G} W_{\gamma_1,\gamma_2}$ with G-action covering

$$(\gamma_1, \gamma_2) \xrightarrow{h} (h\gamma_1 h^{-1}, h\gamma_2 h^{-1})$$

to the G-graded vector space with homogeneous subspaces $\bigoplus_{\gamma_1,\gamma_2\in G,\, \gamma_1\gamma_2=\gamma} W_{\gamma_1,\gamma_2}$ and component-wise G-action. To see that this prescription gives the adjoint functor, we notice that the required isomorphism

$$\mathrm{Hom}(m^*(V), W) \cong \mathrm{Hom}(V, F(W))$$

for $V \in [G/\!/G, \mathcal{V}ect(\mathbb{K})]$ comes from the fact that the vector spaces on either side consist of families of maps

$$f_{\gamma_1,\gamma_2}: \quad V_{\gamma_1\gamma_2} \longrightarrow W_{\gamma_1,\gamma_2}$$

that are compatible with the G-action.

(2) We deal with a system of manifolds and embeddings of the general form (5.17). For the case at hand, this is indicated in the following picture, which for better visibility is shown in a rotated view:

(5.36)

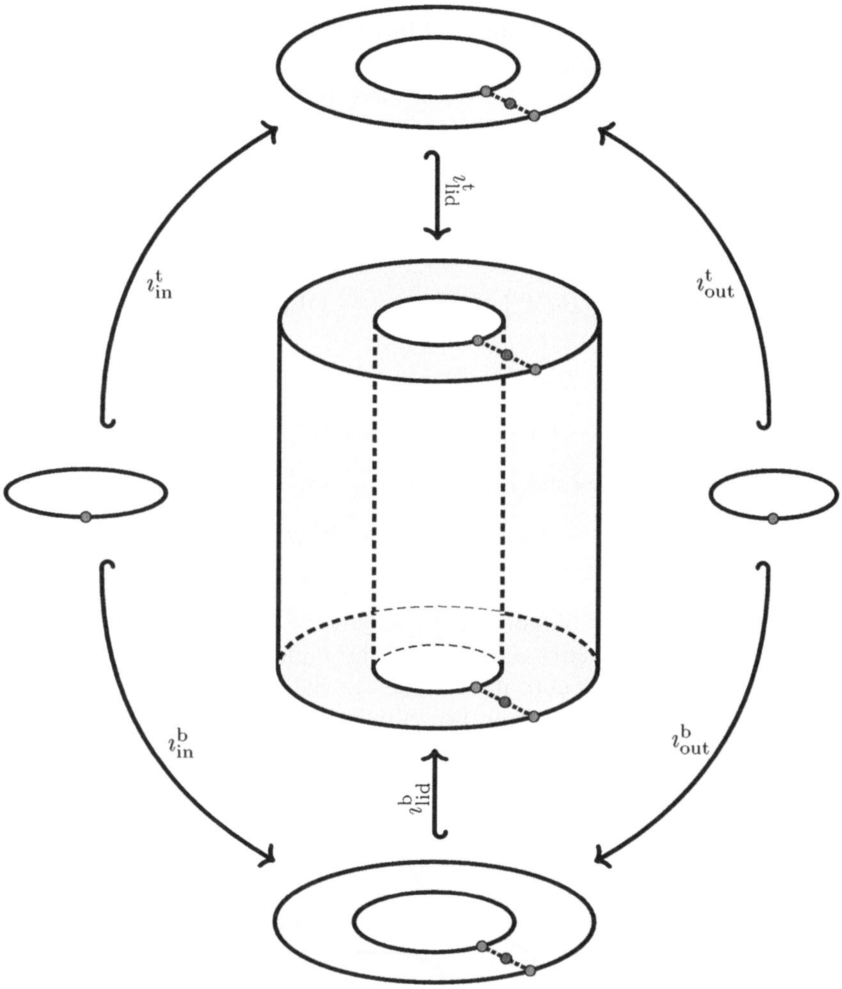

The one-manifolds on the left and right of this picture are copies of the standard model $\{|z|=1\}\subset\mathbb{C}$ of the circle S^1 for which we select $-\mathrm{i}\in\mathbb{C}$ as a base point, as already done in the proof of Proposition 4.49. The two-manifolds A_{t} and A_{b} at the top and bottom of the picture are copies of the standard model A of an annulus, which we take to be the subset $\{1/2\le|z|\le 1\}\subset\mathbb{C}$ of the complex plane, i.e., the unit disk in $\mathbb{C}$ with a smaller disk removed and with base point $p_A=-3\,\mathrm{i}/4\in\mathbb{C}$. We regard the inner boundary circle of A as incoming and the outer one as outgoing. Thus schematically the maps $\imath_{\mathrm{in}}^{\mathrm{b}}$ and $\imath_{\mathrm{out}}^{\mathrm{b}}$ to the bottom lid look like

(5.37)

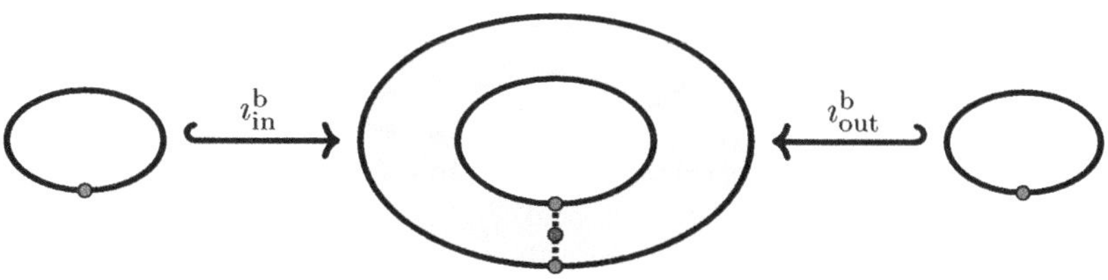

and analogously for the maps $\imath_{\mathrm{out}}^{\mathrm{t}}$ and $\imath_{\mathrm{in}}^{\mathrm{t}}$ to the top lid.

In order to express these embeddings in terms of group elements we work with groupoids $\mathcal{B}un_G^{\mathrm{pt}}$ of pointed bundle as introduced in Definition 1.66. Thus we consider functors $\mathcal{B}un^{\mathrm{pt}}(A,p_A) \to \mathcal{B}un^{\mathrm{pt}}(S^1,p_{\mathrm{out}})$ and $\mathcal{B}un^{\mathrm{pt}}(A,p_A) \to \mathcal{B}un^{\mathrm{pt}}(S^1,p_{\mathrm{in}})$, with $p_{\mathrm{out}} = -\mathrm{i}$ and $p_{\mathrm{in}} = -\mathrm{i}/2 \in \mathbb{C}$ the base points of the incoming and outgoing boundary circle of the annulus A, respectively. For specifying these we must prescribe a map between the chosen reference points in the fibers over p_A and over $p_{\mathrm{out/in}}$. This is achieved via parallel transport along the auxiliary paths a_{out} and a_{in} that are indicated as dashed lines in the picture (5.37). Considering holonomies we then get a commutative diagram

$$\begin{array}{ccc} \mathrm{Obj}(\mathcal{B}un^{\mathrm{pt}}(A,p_A)) & \longrightarrow & \mathrm{Obj}(\mathcal{B}un^{\mathrm{pt}}(S^1,p_{\mathrm{in}})) \\ \Big\downarrow{\scriptstyle \mathrm{hol}} & & \Big\downarrow{\scriptstyle \mathrm{hol}} \\ \mathrm{Hom}(\pi_1(A,p_A),G) & \xrightarrow{\cong} & \mathrm{Hom}(\pi_1(S^1,p_{\mathrm{in}}),G), \end{array}$$

where the map in the second row reads

$$\begin{aligned} \pi_1(A,p_A) &\xrightarrow{\cong} \pi_1(S^1,p_{\mathrm{in}})\,, \\ [s] &\longmapsto [a_{\mathrm{in}} \circ s \circ a_{\mathrm{in}}^{-1}]\,, \end{aligned}$$

and analogously for the outgoing boundary circle.

Consider now the three-manifold in the middle of the picture (5.36); we call it the "twist-tin" and denote it by X_{tw}. As the base point of X_{tw} we choose the base point $\imath_{\mathrm{lid}}^{\mathrm{b}}(A_{\mathrm{b}})$ of its bottom lid. Similarly as in the case of the braid-tin X_{br} (5.26) we parametrize X_{tw} as $A \times [0,1]$ with A the standard model for the annulus. We specify the tubes formed by the locus of the boundary circles at any $t \in [0,1]$ as the embedding $\frac{1}{2}\,\mathrm{e}^{\mathrm{i}\,\alpha} \mapsto (\frac{1}{2}\,\mathrm{e}^{\mathrm{i}\,\alpha},t)$ with $\alpha \in [0,2\pi)$ for the incoming circle and as $\mathrm{e}^{\mathrm{i}\,\alpha} \mapsto (\mathrm{e}^{\mathrm{i}\,\alpha+2\,\mathrm{i}\,\pi t},t)$ with $\alpha \in [0,2\pi)$ for the outgoing one. This is illustrated in the following picture:

(5.38) 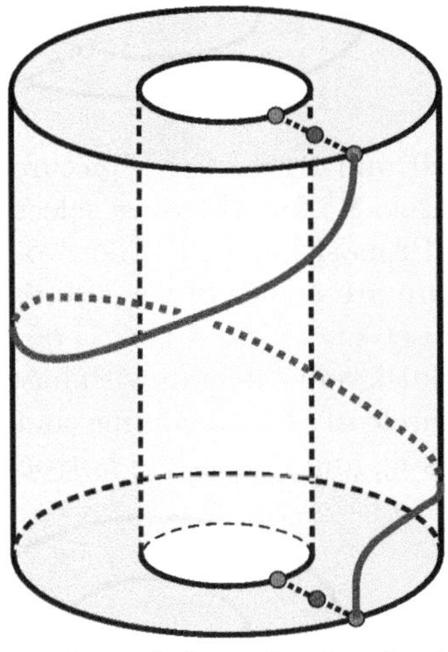

Via the given parametrization of the tube for the incoming boundary circle of the annulus, the two maps $\imath_{\mathrm{lid}}^{\mathrm{b}} \circ \imath_{\mathrm{in}}^{\mathrm{b}}$ and $\imath_{\mathrm{lid}}^{\mathrm{t}} \circ \imath_{\mathrm{in}}^{\mathrm{t}}$ are homotopic. As seen in Example 1.56, this implies that for any G-bundle P_G on X_{tw} we have a natural isomorphism

$$(\imath_{\mathrm{lid}}^{\mathrm{b}} \circ \imath_{\mathrm{in}}^{\mathrm{b}})^* P_G \xrightarrow{\cong} (\imath_{\mathrm{lid}}^{\mathrm{t}} \circ \imath_{\mathrm{in}}^{\mathrm{t}})^* P_G\,.$$

By taking bundle groupoids we then get the diagram

$$
\begin{array}{ccccc}
 & & \mathcal{B}un_G(A_{\mathrm{t}}) & & \\
 & \swarrow^{(\imath_{\mathrm{in}}^{\mathrm{t}})^*} & \uparrow{\scriptstyle (\imath_{\mathrm{lid}}^{\mathrm{t}})^*} & \searrow^{(\imath_{\mathrm{out}}^{\mathrm{t}})^*} & \\
\mathcal{B}un_G(S^1) & \Longleftarrow & \mathcal{B}un_G(X_{\mathrm{tw}}) & \Longrightarrow & \mathcal{B}un_G(S^1) \\
 & \nwarrow_{(\imath_{\mathrm{in}}^{\mathrm{b}})^*} & \downarrow{\scriptstyle (\imath_{\mathrm{lid}}^{\mathrm{b}})^*} & \nearrow_{(\imath_{\mathrm{out}}^{\mathrm{b}})^*} & \\
 & & \mathcal{B}un_G(A_{\mathrm{b}}) & &
\end{array}
$$

of groupoids and functors, with natural transformations arising from homotopies as fillers. Our strategy is now to work with the equivalent groupoids $\mathcal{B}un_G^{\mathrm{pt}}$ and then reduce to 'small' realizations, whereby we deal with the diagram

$$
\begin{array}{ccccc}
 & & G/\!/G & & \\
 & \swarrow^{\mathrm{id}} & \uparrow{\scriptstyle F_{\mathrm{lid}}^{\mathrm{t}}} & \searrow^{\mathrm{id}} & \\
G/\!/G & \Longleftarrow & G/\!/G & \Longrightarrow & G/\!/G. \\
 & \nwarrow_{\mathrm{id}} & \downarrow{\scriptstyle F_{\mathrm{lid}}^{\mathrm{b}}} & \nearrow_{\mathrm{id}} & \\
 & & G/\!/G & &
\end{array}
\tag{5.39}
$$

To find the functors $F_{\mathrm{lid}}^{\mathrm{t}}$ and $F_{\mathrm{lid}}^{\mathrm{b}}$ and the natural transformations that furnish the fillers in this diagram we proceed as follows. On the top and bottom lid we use parallel transport along auxiliary paths to specify the reference points in the fibers. In this way we get the identity functors on $G/\!/G$ as indicated in (5.39).

Next we relate the twist-tin X_{tw} to the annuli at its lids in the following way. At the bottom, we select the same base point p_A^{b} as for the annulus and, for a given choice of G-bundle P, the same reference point in the fiber $P_{p_A^{\mathrm{b}}}$. At the top we take again the base point p_A^{t} of the annulus and perform parallel transport along the parametrized tubes, properly accounting for the principle of globularity (as expounded in Remark 5.16(3)). This way we specify two points $q_{\mathrm{in}}, q_{\mathrm{out}} \in P_{p_A^{\mathrm{t}}}$ in the same fiber. It follows that $F_{\mathrm{lid}}^{\mathrm{t}}$ and $F_{\mathrm{lid}}^{\mathrm{b}}$ are the identity functor on $G/\!/G$. Moreover, if we use q_{in} as a reference point in the fiber, then we see that the filler of the left quadrangle in (5.39) is the identity natural transformation. On the other hand, the points q_{in} and q_{out} are different; we have $q_{\mathrm{out}} = g.q_{\mathrm{in}}$ with g the holonomy of the bundle P_G around the noncontractible closed path that is obtained by concatenating the path on the outer tube with the inverse of the path on the inner tube (and with the appropriate auxiliary paths on the lids). Accordingly, the natural transformation $\alpha\colon \mathrm{id} \Rightarrow \mathrm{id}$ that is the filler of the right quadrangle in (5.39) has components

$$\alpha_\gamma : \quad \mathrm{id}(\gamma) = \gamma \xrightarrow{\ \gamma\ } \gamma\gamma\gamma^{-1} = \gamma = \mathrm{id}(\gamma)$$

for $\gamma \in G$. After linearization, this induces the balancing in the claimed form.

(3) The manifolds and embeddings relevant for the braiding are indicated in the following picture:

(5.40)

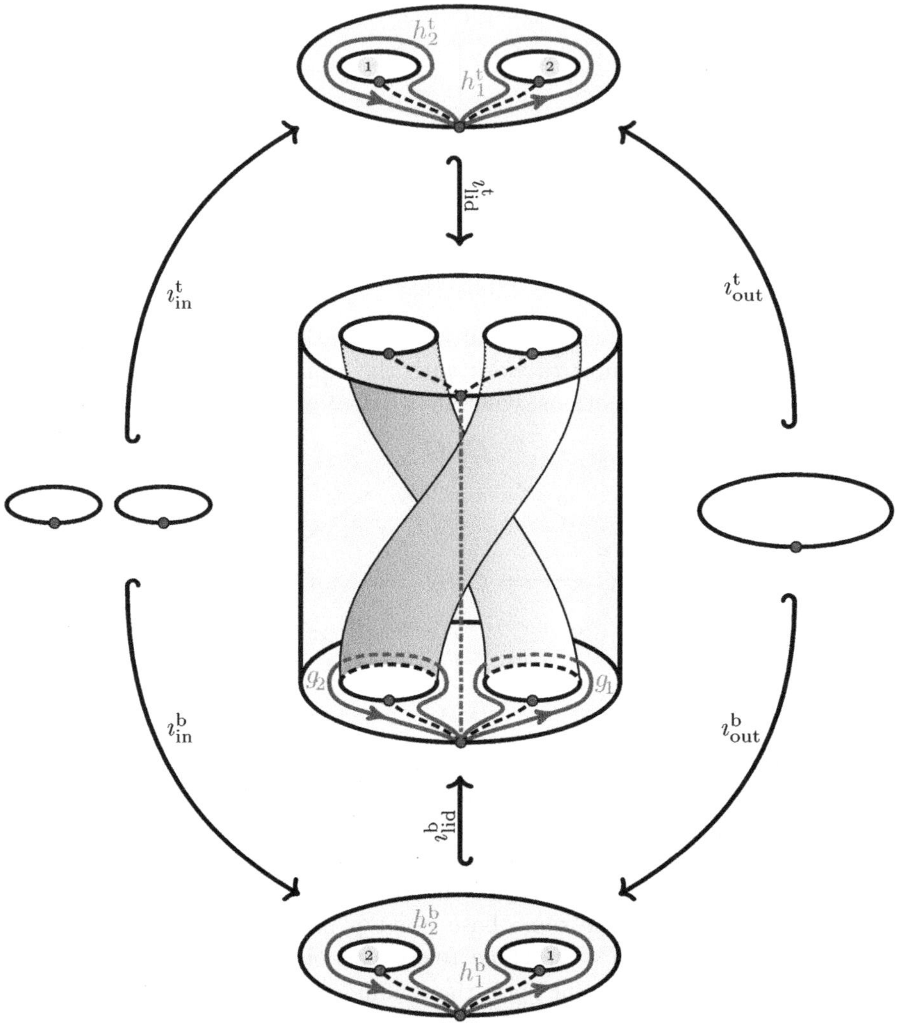

Here the two-manifolds T_{t} and T_{b} at the top and bottom of the picture are copies of the standard model of the pair of pants shown in (4.52), including the choice of base points of their boundary circles. But instead of taking over the more symmetric choice of base point of T made in (4.52) we select the base point for the fundamental groups of T_{t} and T_{b} to coincide with the base points of the respective outgoing circles, as this will slightly simplify the discussion below. We denote the generators of $\pi_1(T_{\mathrm{t}}, p_0^{\mathrm{t}})$ and $\pi_1(T_{\mathrm{b}}, p_0^{\mathrm{b}})$ by $h_{1,2}^{\mathrm{t}}$ (top) and by $h_{1,2}^{\mathrm{b}}$ (bottom), respectively. We refer to the three-manifold with corners in the middle of the picture again as the braid-tin X_{br} and choose its base point p_X to coincide with the base point of the outgoing boundary circle of the bottom lid. The fundamental group $\pi_1(X_{\mathrm{br}}, p_X)$ of the braid-tin is generated by the paths that we denote by g_1 and g_2.

The two maps $\imath_{\mathrm{out}}^{\mathrm{b}}$ and $\imath_{\mathrm{out}}^{\mathrm{t}}$ on the right part of the picture (5.40) embed the standard circle to the outgoing boundary of the bottom and top lid, respectively, preserving the base point. The two maps $\imath_{\mathrm{lid}}^{\mathrm{b}}$ and $\imath_{\mathrm{lid}}^{\mathrm{t}}$ in the middle part embed the respective copies of (4.52) as the bottom and top lids of the tin. The two maps $\imath^{\mathrm{b}}_{\mathrm{in}}$ and $\imath^{\mathrm{t}}_{\mathrm{in}}$ on the left part of (5.40) embed the disjoint union $S^1 \sqcup S^1$ (i.e., an ordered pair of standard circles, which we indicate by attaching labels 1 and 2 to them), to the two incoming boundary circles of the top and bottom lid, in such a way that base points are preserved and that the total map to X_{br} is in agreement with the principle of globularity. Thus $\imath^{\mathrm{b}}_{\mathrm{in}}$ preserves the labels of the circles, while $\imath^{\mathrm{t}}_{\mathrm{in}}$ exchanges them.

Consider now G-bundles over X_{br} and over the embedded lids and boundary circles and describe them in terms of holonomies, leading to finite groupoids. Restriction of bundles then furnishes the following span of spans of groupoids:

$$
\begin{array}{ccccc}
 & & (G\times G)/\!/G & & \\
 & \swarrow^{F^{\mathrm{t}}_{\mathrm{in}}} & \uparrow F^{\mathrm{t}}_{\mathrm{lid}} & \searrow^{F^{\mathrm{t}}_{\mathrm{out}}} & \\
G/\!/G\times G/\!/G & \Longleftarrow & (G\times G)/\!/G & \Longrightarrow & G/\!/G. \\
 & \nwarrow_{F^{\mathrm{b}}_{\mathrm{in}}} & \downarrow F^{\mathrm{b}}_{\mathrm{lid}} & \nearrow_{F^{\mathrm{b}}_{\mathrm{out}}} & \\
 & & (G\times G)/\!/G & &
\end{array}
\tag{5.41}
$$

Here in the left and right columns G acts by the adjoint action, and in the middle column by the diagonal adjoint action. The functors in this diagram are determined by the effect of pullback of bundles on holonomies. Our task is to obtain these functors and the natural transformations that are the fillers in the two squares of (5.41) explicitly. For $F^{\mathrm{b}}_{\mathrm{lid}}$ this is immediate: owing to the identification of the base points of the bottom lid and of X_{br} it is the identity functor. The two functors $F^{\mathrm{b}}_{\mathrm{out}}, F^{\mathrm{t}}_{\mathrm{out}}\colon (G\times G)/\!/G \longrightarrow G/\!/G$ coincide; they map objects as $(\gamma_1,\gamma_2)\mapsto\gamma_1\gamma_2$ and map a morphism $(\gamma_1,\gamma_2)\xrightarrow{h}(h\gamma_1h^{-1},h\gamma_2h^{-1})$ to

$$
F^{\mathrm{b,t}}_{\mathrm{out}}(h):\quad \gamma_1\gamma_2 \longrightarrow (h\gamma_1h^{-1})\,(h\gamma_2h^{-1}) = h\gamma_1\gamma_2h^{-1}.
$$

Next, since $\imath^{\mathrm{b}}_{\mathrm{in}}$ preserves the numbering of the circles, $F^{\mathrm{b}}_{\mathrm{in}}$ maps objects trivially as $(\gamma_1,\gamma_2)\mapsto(\gamma_1,\gamma_2)$, while on morphisms it is the diagonal map. In contrast, since $\imath^{\mathrm{t}}_{\mathrm{in}}$ reverses the numbering, $F^{\mathrm{t}}_{\mathrm{in}}$ maps objects as $(\gamma_1,\gamma_2)\mapsto(\gamma_2,\gamma_1)$ and maps a morphism $h\colon(\gamma_1,\gamma_2)\longrightarrow(h\gamma_1h^{-1},h\gamma_2h^{-1})$ to a morphism

$$
F^{\mathrm{t}}_{\mathrm{in}}(h):\quad (\gamma_1,\gamma_2)\longrightarrow(h\gamma_2h^{-1},h\gamma_1h^{-1})\,.
$$

Finally, for computing $F^{\mathrm{t}}_{\mathrm{lid}}$ we must relate the two generators of the fundamental group $\pi_1(X_{\mathrm{br}},p_X)$ of the tin to paths in the top lid. This is done by concatenating paths in the following manner: starting at the base point p^{t}_0 of the top lid, follow the path to p_X that in the picture (5.40) is drawn as a dashed-dotted line, then along either of the two generators of $\pi_1(X_{\mathrm{br}},p_X)$, and then back along the dashed-dotted path to p_0^{t}. The so obtained path can be contracted within X_{br} to a closed path in the top lid; the resulting images of the two generators g_1 and g_2 of $\pi_1(X_{\mathrm{br}}, p_X)$ look as follows:

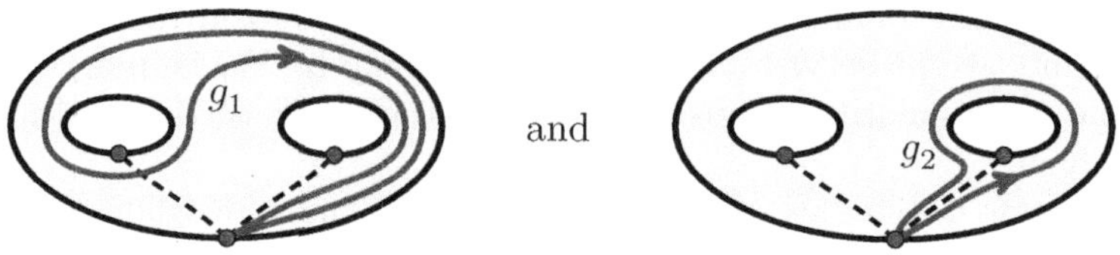

The path obtained as the image of g_1 can be expressed as the concatenation of a path running counter-clockwise around the right circle, followed by a path counter-clockwise around the left circle, and then clockwise around the right circle. This shows that $F_{\mathrm{lid}}^{\mathrm{t}}$ maps objects according to $(\gamma_1, \gamma_2) \mapsto (\gamma_1\gamma_2\gamma_1^{-1}, \gamma_1)$. We can combine this information into the diagram

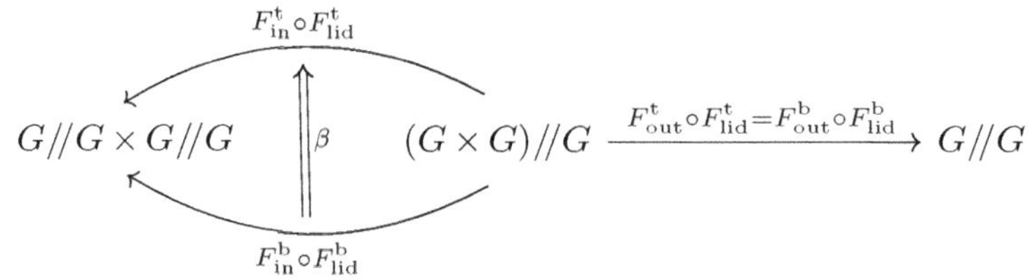

of functors between groupoids. Here on the left part we need a natural transformation β as a filler between the noncoinciding functors $F_{\mathrm{in}}^{\mathrm{b}} \circ F_{\mathrm{lid}}^{\mathrm{b}}$ and $F_{\mathrm{in}}^{\mathrm{t}} \circ F_{\mathrm{lid}}^{\mathrm{t}}$. As can be read off the mapping prescriptions for these functors, the components of β are morphisms

$$\begin{aligned}\beta_{(\gamma_1,\gamma_2)} : \quad (\gamma_1, \gamma_2) = F_{\mathrm{in}}^{\mathrm{b}} \circ F_{\mathrm{lid}}^{\mathrm{b}}(\gamma_1, \gamma_2) \\ \longrightarrow F_{\mathrm{in}}^{\mathrm{t}} \circ F_{\mathrm{lid}}^{\mathrm{t}}(\gamma_1, \gamma_2) = (\gamma_1^{-1}\gamma_2\gamma_1, \gamma_1)\end{aligned}$$

in $G//G \times G//G$. An analysis similar to the one for the case of the balancing reveals that these components are the morphisms

$$\beta_{(\gamma_1,\gamma_2)} = (\gamma_1, e) \,. \tag{5.42}$$

Finally we linearize. The functor

$$(F_{\mathrm{in}}^{\mathrm{b}} \circ F_{\mathrm{lid}}^{\mathrm{b}})^* : \quad [G//G \times G//G, \mathcal{V}ect(\mathbb{K})] \longrightarrow [(G{\times}G)//G, \mathcal{V}ect(\mathbb{K})]$$

maps the $G{\times}G$-graded vector space $V = \bigoplus_{g_1,g_2 \in G} V_{g_1,g_2}$ endowed with the $G{\times}G$-action covering the adjoint action $(h_1, h_2).V_{g_1,g_2} \subset V_{h_1g_1h_1^{-1},h_2g_2h_2^{-1}}$ to the same $G{\times}G$-graded vector space endowed with a G-action $h.V_{g_1,g_2} \subset V_{hg_1h^{-1},hg_2h^{-1}}$. The functor

$$(F_{\mathrm{in}}^{\mathrm{t}} \circ F_{\mathrm{lid}}^{\mathrm{t}})^* : \quad [G//G \times G//G, \mathcal{V}ect(\mathbb{K})] \longrightarrow [(G{\times}G)//G, \mathcal{V}ect(\mathbb{K})]$$

changes the grading in such a way that the component of degree (g_1, g_2) is $V_{g_2g_1g_2^{-1},g_2}$. Hence on the homogeneous subspace V_{g_1,g_2} the natural transformation

$$\beta^* : \quad (F_{\mathrm{in}}^{\mathrm{b}} \circ F_{\mathrm{lid}}^{\mathrm{b}})^* \longrightarrow (F_{\mathrm{in}}^{\mathrm{t}} \circ F_{\mathrm{lid}}^{\mathrm{t}})^*$$

is given by the g_2-action on the first argument. Now we already know that the composite

$$\otimes := (F_{\mathrm{out}}^{\mathrm{b}} \circ F_{\mathrm{lid}}^{\mathrm{b}})_* \circ (F_{\mathrm{in}}^{\mathrm{b}} \circ F_{\mathrm{lid}}^{\mathrm{b}})^* : \quad \mathcal{C}_{\mathbb{K}}(G) \boxtimes \mathcal{C}_{\mathbb{K}}(G) \longrightarrow \mathcal{C}_{\mathbb{K}}(G)$$

is the tensor product. Likewise,

$$\otimes^{\mathrm{opp}} := (F^{\mathrm{t}}_{\mathrm{out}} \circ F^{\mathrm{t}}_{\mathrm{lid}})_* \circ (F^{\mathrm{t}}_{\mathrm{in}} \circ F^{\mathrm{t}}_{\mathrm{lid}})^* : \quad \mathcal{C}_{\mathbb{K}}(G) \times \mathcal{C}_{\mathbb{K}}(G) \longrightarrow \mathcal{C}_{\mathbb{K}}(G)$$

is the opposite tensor product. Thus the braiding we are looking for is the horizontal composition

$$\mathrm{id}_{(F^{\mathrm{b}}_{\mathrm{out}} \circ F^{\mathrm{b}}_{\mathrm{lid}})} \star \beta^* : \quad \otimes \longrightarrow \otimes^{\mathrm{opp}}$$

of natural transformations. From the result for β^* we find that its component

$$\big(\mathrm{id}_{(F^{\mathrm{b}}_{\mathrm{out}} \circ F^{\mathrm{b}}_{\mathrm{lid}})} \star \beta^*\big)_{(\gamma_1,\gamma_2)} = (F^{\mathrm{b}}_{\mathrm{out}} \circ F^{\mathrm{b}}_{\mathrm{lid}})(\beta^*_{(\gamma_1,\gamma_2)}) :$$

$$\bigoplus_{\gamma\in G} \bigoplus_{\substack{\gamma_1,\gamma_2\in G\\ \gamma_1\gamma_2=\gamma}} V_{\gamma_1} \otimes W_{\gamma_2} \longrightarrow \bigoplus_{\gamma\in G} \bigoplus_{\substack{\gamma_1,\gamma_2\in G\\ \gamma_1\gamma_2=\gamma}} W_{\gamma_2} \otimes V_{\gamma_1}$$

at (γ_1, γ_2) is given by

$$v \otimes w \mapsto \gamma_1 . w \otimes v$$

for homogeneous vectors $v \otimes w \in V_{\gamma_1} \otimes W_{\gamma_2}$. (Note that $\gamma_1 . w \otimes v$ has the same degree $(\gamma_1\gamma_2\gamma_1^{-1})\,\gamma_1 = \gamma_1\gamma_2$ as $v \otimes w$, as required.) □

EXERCISE 5.43. Complete the derivation of the formula (5.42) for the natural transformation that determines the braiding on $\mathcal{C}_{\mathbb{K}}(G)$.

For a finite group G, we denote by $\mathbb{K}(G)$ the commutative algebra that is obtained by endowing the vector space $\mathrm{Map}(G,\mathbb{K})$ of $\mathbb{K}$-valued functions on G with the pointwise product of functions, i.e., $(f_1\, f_2)(g) := f_1(g)\, f_2(g)$. (Recall that $\mathrm{Map}(G,\mathbb{K})$ can also be endowed with the convolution product (4.44), which is non-commutative if G is not abelian.) The vector space $\mathrm{Map}(G,\mathbb{K})$ has a basis $(\delta_g)_{g\in G}$ consisting of the group delta functions (4.45) that are supported in a single group element. These form a complete set of orthogonal idempotents under the pointwise product:

$$\delta_g\, \delta_h = \delta_{g,h}\, \delta_h \qquad \text{and} \qquad \sum_{g\in G} \delta_g = 1\,. \tag{5.44}$$

Further, as in Definition 1.15 we denote by $\mathbb{K}[G]$ the group algebra of G, with basis $\{\beta_g \,|\, g \in G\}$. The tensor product vector space

$$\mathcal{D}(G) := \mathbb{K}(G) \otimes_{\mathbb{K}} \mathbb{K}[G]$$

carries a multiplication given by

$$(\delta_a \otimes \beta_b)\,(\delta_c \otimes \beta_d) = \delta_a\, \delta_{bcb^{-1}} \otimes \beta_b\, \beta_d \quad \text{for all } a,b,c,d \in G\,.$$

This multiplication makes $\mathcal{D}(G)$ into a unital associative algebra, which is called the *Drinfeld double* of G. By definition, the multiplication satisfies in particular

$$(1 \otimes \beta_g)\,(\delta_h \otimes \beta_e) = \delta_{ghg^{-1}} \otimes \beta_g \tag{5.45}$$

for $g, h \in G$; this is called the *straightening formula*.

PROPOSITION 5.46. The category $\mathcal{D}(G)$-mod of finite-dimensional modules over the Drinfeld double $\mathcal{D}(G)$ is equivalent as an abelian category to the category $\mathcal{C}_{\mathbb{K}}(G)$ extracted from the extended Dijkgraaf-Witten theory.

PROOF. By restriction to the subalgebras $\mathbb{K}(G)$ and $\mathbb{K}[G]$, a finite-dimensional $\mathcal{D}(G)$-module (V, ρ) has the structure of a $\mathbb{K}(G)$-module and of a $\mathbb{K}[G]$-module. As a consequence of the completeness relation (5.44) for the group delta functions $(\delta_a)_{a \in G}$ in $\mathbb{K}(G)$ we have a direct sum decomposition $V = \bigoplus_{g \in G} \mathrm{Im}\,(\delta_g)$, i.e., $V = \bigoplus_{g \in G} V_g$ is a graded vector space, as in Proposition 5.33. In particular, on a homogeneous vector v_g of degree $g \in G$ we have

$$(\delta_a \otimes \beta_e).v_g = \delta_{a,g}\, v_g\,.$$

Also, by applying the straightening formula (5.45) to vectors in V_h one sees that $g.V_h \subseteq V_{ghg^{-1}}$, as in (5.34). This implies that $(\delta_a \otimes \beta_g).v_h = \delta_{a,ghg^{-1}}\, g.v_h$ for $v_h \in V_h$. We leave the full details of the proof to the reader. □

For a finite group G, a $\mathbb{K}[G]$-comodule structure on a vector space V amounts to a $\mathbb{K}[G]^*$-module structure. Using that $\mathbb{K}[G]^* = \mathbb{K}(G)$, a $\mathcal{D}(G)$-module can therefore alternatively be described as a vector space that is endowed with a module and a comodule structure together with an appropriate compatibility condition. Such a structure is called a *Yetter-Drinfeld module.* For details about this description we refer to [**Kas**, Ch. IX.5], where Yetter-Drinfeld modules are called crossed bimodules.

REMARK 5.47. The category of $\mathcal{D}(G)$-modules is in particular braided. This is in fact a special case of the following much more general result: To any monoidal category $\mathcal{C}$ one can associate a braided monoidal category $\mathcal{Z}(\mathcal{C})$, called the *Drinfeld center* of $\mathcal{C}$.

An object of $\mathcal{Z}(\mathcal{C})$ is a pair consisting of an object $X \in \mathcal{C}$ and a *half-braiding* for X. The latter is a natural family $\sigma = (\sigma_Y)_{Y \in \mathcal{C}}$ of morphisms $\sigma_Y : Y \otimes X \longrightarrow X \otimes Y$ such that the hexagon identity

$$(\sigma_Y \otimes \mathrm{id}_{Y'}) \circ a^{-1}_{Y,X,Y'} \circ (\mathrm{id}_Y \otimes \sigma_{Y'}) = a^{-1}_{Y,X,Y'} \circ \sigma_{Y \otimes Y'} \circ a^{-1}_{Y,Y',X}$$

holds for all $Y, Y' \in \mathbb{C}$. The morphisms $\mathrm{Hom}_{\mathcal{Z}(\mathcal{C})}((X, \sigma), (X', \sigma'))$ are those morphisms $f \colon X \longrightarrow X'$ in $\mathcal{C}$ that satisfy $(f \otimes \mathrm{id}_Y) \circ \sigma_Y = \sigma'_Y \circ (\mathrm{id}_Y \otimes f)$ for all $Y \in \mathcal{C}$. Notably, this construction works without assuming $\mathcal{C}$ to be braided, and thus even for monoidal categories $\mathcal{C}$ that do not admit any braiding at all.

This construction generalizes further to the *relative Drinfeld center* $\mathcal{Z}_{\mathcal{C}}(\mathcal{M})$ of a module category $\mathcal{M}$ over a monoidal category $\mathcal{C}$, which is equivalent to the category of $\mathcal{C}$-bimodule functors from $\mathcal{C}$ to $\mathcal{M}$ [**GelNN**].

REMARK 5.48. The functor

$$\begin{aligned} G\text{-}\mathcal{R}ep &\longrightarrow \mathcal{D}(G)\text{-mod}\,, \\ V &\longmapsto V_e \end{aligned}$$

is monoidal and fully faithful. Thus the symmetric monoidal category G-$\mathcal{R}ep$ is a full monoidal subcategory of $\mathcal{D}(G)$-mod. In fact, the objects of $\mathcal{D}(G)$-mod are pairs consisting of those objects of the category G-$\mathcal{R}ep$ that admit braidings and all possible braidings on them. Alternatively, one can describe them also as pairs consisting of objects of the category $\mathcal{V}ect_G(\mathbb{K})$ together with all possible braidings.

5.6. 3-2-1-dimensional TFTs and modular categories

According to Proposition 5.29, every extended three-dimensional topological field theory produces, by evaluation on the circle, a balanced braided category. In other words, the evaluation on the circle provides a map from extended three-dimensional topological field theories to balanced braided categories. This raises the natural question whether we can, analogously as for two-dimensional topological field theories, use this map for achieving a classification. Indeed, we will see that the evaluation on the circle provides an equivalence from extended three-dimensional topological field theories to balanced braided categories *with some additional properties.*

In addition to the notion of a ribbon category that we provided in Definition 5.28 we need

DEFINITION 5.49.

(1) A *finite category* over a field $\mathbb{K}$ is a $\mathbb{K}$-linear category that is linearly equivalent to the category of finite-dimensional modules over a finite-dimensional $\mathbb{K}$-algebra.
(No such equivalence is selected.)
(2) A *finite tensor category* over $\mathbb{K}$ is a finite category with rigid tensor product and simple monoidal unit $\mathbf{1}$ (i.e., $\mathbf{1}$ has exactly two nonisomorphic subobjects, namely $\mathbf{1}$ itself and 0).
(3) A *finite ribbon category* is a finite tensor category with braiding and balancing that obey the ribbon condition from Definition 5.28(2).
(4) A *fusion category* is a semisimple finite tensor category.

REMARK 5.50. Requiring the monoidal unit to be simple is unimportant: Any semisimple braided monoidal category with possibly nonsimple monoidal unit splits into the direct sum of semisimple braided monoidal categories with simple unit [**BDSV**, Lemma 5.3].

The next two exercises show how the categorical structures just introduced are realized on representation categories of Hopf algebras.

EXERCISE 5.51. Let A be a finite-dimensional Hopf algebra over a field $\mathbb{K}$. Using Theorem 2.84, conclude that the category A-mod of finite-dimensional A-modules over $\mathbb{K}$ comes with the structure of a finite tensor category.

EXERCISE 5.52. Let A be a finite-dimensional Hopf algebra that is quasi-triangular with R-matrix R. By combining Exercises 5.51 and 2.107, show that A-mod comes with the structure of a finite braided tensor category.

One defines a *ribbon element* to be an invertible central element $\nu \in A$ satisfying

$$\Delta(\nu) = (R_{21}R)^{-1} \cdot (\nu \otimes \nu), \quad \varepsilon(\nu) = 1 \quad \text{and} \quad \mathrm{S}(\nu) = \nu .$$

Prove that any ribbon element of A yields a ribbon structure on the category A-mod whose component at a finite-dimensional A-module V is given by

$$\begin{aligned} \theta_V : \quad & V \longrightarrow V , \\ & v \longmapsto \nu^{-1}.v . \end{aligned}$$

We next need to characterize the braiding that appears on categories arising from the evaluation of a topological field theory on a circle.

DEFINITION 5.53.

(1) A *transparent* object in a braided monoidal category $\mathcal{C}$ is an object $X \in \mathcal{C}$ such that

$$c_{Y,X} \circ c_{X,Y} = \mathrm{id}_{X\otimes Y}$$

for every $Y \in \mathcal{C}$.

(Thus the monoidal unit $\mathbf{1}$ is a transparent object.)

(2) A braiding on a finite tensor category is called *nondegenerate* if every transparent object is isomorphic to a finite direct sum $\mathbf{1} \oplus \cdots \oplus \mathbf{1}$ of copies of the monoidal unit.

(3) A *modular category*, also called *modular tensor category*, is a finite ribbon category with a nondegenerate braiding.

(4) A semisimple modular category is also called a *modular fusion category*.

REMARKS 5.54.

(1) A ribbon category has a distinguished pivotal structure. This pivotal structure is spherical, so that in any ribbon category the left trace and right trace coincide.

(2) Given a ribbon fusion category $\mathcal{C}$ and a set $\{U_i\}_{i\in I}$ of representatives of its isomorphism classes of simple objects, one defines two square matrices S and T as follows: S is the symmetric matrix with entries

$$S_{i,j} := \mathrm{tr}\left(c_{U_j,U_i} \circ c_{U_j,U_i}\right),$$

while T is the diagonal matrix whose diagonal entries are the twist eigenvalues of the objects U_i, i.e., $T_{i,i} = \vartheta_i$ with $\theta_{U_i} = \vartheta_i \, \mathrm{id}_{U_i}$. The matrix S is invertible if and only if the category $\mathcal{C}$ is modular.

(3) The two matrices S and T together are often referred to as the *modular data* of $\mathcal{C}$. These data can be explicitly computed for some classes of modular fusion categories and have therefore attracted a lot of attention. However, they do not determine a modular tensor category: there are inequivalent modular tensor categories having the same modular data [**MiS**].

(4) One can characterize the nondegeneracy of the braiding alternatively as follows. For a braided finite tensor category $\mathcal{C}$ denote by $\mathcal{C}^{\mathrm{rev}}$ the same monoidal category as $\mathcal{C}$, but with inverted braiding. Recall the notion of Deligne tensor product $\boxtimes$ from page 124; the Deligne product $\mathcal{C} \boxtimes \mathcal{C}^{\mathrm{rev}}$ comes with a distinguished braided monoidal functor to the Drinfeld center $\mathcal{Z}(\mathcal{C})$. The braiding is nondegenerate if and only if this functor is an equivalence of categories. Further characterizations of nondegenerately braided categories are given in [**Shi**].

We can now refine Proposition 5.29:

THEOREM 5.55 ([**BDSV**, Prop. 4.8, Thm. 4.9 & Cor. 5.15]). The balanced braided category obtained by evaluating a 3-2-1-dimensional topological field theory on the circle is a modular fusion category. In particular, it is rigid.

EXAMPLE 5.56. The Drinfeld double $\mathcal{D}(G)$ of a finite group G introduced in Section 5.5 can be endowed with the structure of a quasi-triangular Hopf algebra as follows. In terms of the bases $\{\delta_g\}$ of the function algebra $\mathbb{K}(G)$ and $\{\beta_g\}$ of the group algebra $\mathbb{K}[G]$ the coproduct, counit and antipode are given by

$$\Delta(\delta_a \otimes \beta_b) := \sum_{c\in G} \delta_c \otimes \beta_b \otimes \delta_{c^{-1}a} \otimes \beta_b \,, \tag{5.57}$$

and

$$\varepsilon(\delta_a \otimes \beta_b) := \delta_{a,e} \qquad \text{and} \qquad \mathrm{s}(\delta_a \otimes \beta_b) := \delta_{b^{-1}a^{-1}b} \otimes \beta_{b^{-1}}$$

for $a, b \in G$. Let us verify that the coproduct (5.57) matches the tensor product derived geometrically in Proposition 5.35: Using that acting with the orthogonal idempotents δ_h projects to homogeneous components, we obtain

$$\begin{aligned}(\delta_a \otimes \beta_b).(v_g \otimes w_h) &= \sum_{c\in G} (\delta_c \otimes \beta_b).v_g \otimes (\delta_{c^{-1}a} \otimes \beta_b).w_h \\ &= \sum_{c\in G} (b.v_g)_c \otimes (b.w_h)_{c^{-1}h}\end{aligned}$$

for homogeneous vectors $v_g \in V_g$ and $w_h \in W_h$ in $\mathcal{D}(G)$-modules V and W. We thus find a diagonal action of $b \in G$ together with preservation of the total homogeneous degree, in full agreement with Proposition 5.35(1).

Further, by direct computation one checks that

$$R := \sum_{c\in G} \delta_c \otimes \beta_e \otimes 1 \otimes \beta_c$$

with $1 := \sum_{a\in G} \delta_a$ is an R-matrix for the so defined Hopf algebra structure on $\mathcal{D}(G)$, and that

$$\nu := \sum_{c\in G} \delta_c \otimes \beta_{c^{-1}}$$

is a ribbon element. By Exercise 5.52, this makes the category $\mathcal{D}(G)$-mod of finite-dimensional $\mathcal{D}(G)$-modules a finite ribbon category. Also, the braiding can be verified to be nondegenerate, hence this category is modular. By Exercise 2.107 the braiding on the monoidal category $\mathcal{D}(G)$-mod obtained from the R-matrix is given by

$$c^R_{V,W}: \quad v \otimes w \longmapsto R_2.w \otimes R_1.v$$

for v and w vectors in $\mathcal{D}(G)$-modules V and W, respectively. Applying this prescription to homogeneous vectors $v_g \in V_g$ and $w_h \in W_h$ gives

$$\begin{aligned}R_2.w_h \otimes R_1.v_g &= \sum_{c\in G} (1 \otimes \beta_c).w_h \otimes (\delta_c \otimes \beta_e).v_g \\ &= (g.w_h) \otimes v_g \,.\end{aligned}$$

Here the second equality follows because from the second tensorand only the homogeneous component in degree g contributes to the sum. The so obtained braiding is exactly the one derived geometrically in Proposition 5.35(3).

Finally, to determine the twist $\theta(v_g) = \nu^{-1}.v_g$ on a homogeneous element $v_g \in V$ of degree $g \in G$, we compute

$$\nu.v_g = \sum_{c\in G} \delta_c \, c^{-1}.v_g = g^{-1}.v_g \,.$$

Here we use that the element $c^{-1}.v_g$ is homogeneous of degree $c^{-1}g\,c$, which forces $c^{-1}g\,c = c$ and thus selects the term with $c = g$ in the sum. Since the action of g is invertible, we thus get

$$\theta(v_g) = \nu^{-1}.v_g = g.v_g\,.$$

This coincides with the result we derived geometrically in Proposition 5.35(2).

EXERCISE 5.58. The (once-)extended three-dimensional Dijkgraaf-Witten theory Z_G for a finite group G associates to the torus $S^1 \times S^1$ the vector space that is spanned by the simple objects of the category of $\mathcal{C}_{\mathbb{K}}(G)$. (To prove this statement requires concepts that are somewhat beyond what we provide in this text.)

Use this information to show that the number of irreducible representations of the Drinfeld double of G equals the number of commuting triples $(g_1, g_2, g_3) \in G \times G \times G$ divided by the order $|G|$. (See also Theorem 24 in [**Wil**].)

EXAMPLES 5.59. Let us present a few further known classes of modular tensor categories.

(1) Let A be a finite abelian group. Consider the category $\mathcal{V}ect_A(\mathbb{K})$ of A-graded $\mathbb{K}$-vector spaces, with the monoidal structure as described in Example 2.15(5). Recall from Example 2.47 that we can modify the associator

$$a^0_{g_1,g_2,g_3}: \quad (V_{g_1} \otimes V_{g_2}) \otimes V_{g_3} \longrightarrow V_{g_1} \otimes (V_{g_2} \otimes V_{g_3})$$

that is inherited from $\mathcal{V}ect(\mathbb{K})$ by a 3-cocycle $\omega\colon G \times G \times G \to \mathbb{K}^\times$. Validity of the pentagon axiom for $a_{g_1,g_2,g_3} := \omega(g_1, g_2, g_3)\, a^0_{g_1,g_2,g_3}$ is equivalent to the closedness condition (2.48) obeyed by ω.

Similarly, the symmetric braiding

$$c^0_{g_1,g_2}: \quad V_{g_1} \otimes V_{g_2} \to V_{g_2} \otimes V_{g_1}$$

on $\mathcal{V}ect_A(\mathbb{K})$ that is inherited from $\mathcal{V}ect(\mathbb{K})$ can be modified (using that A is abelian) by a function $\gamma\colon G \times G \to \mathbb{K}^\times$. The requirement that

$$c_{g_1,g_2} := \gamma(g_1, g_2)\, c^0_{g_1,g_2}$$

is a braiding, i.e., obeys the two hexagon axioms (see Definition 2.89(2)), leads to polynomial equations for γ, which can easily be worked out (or be looked up in, e.g., [**MooS2**, App. E]).

In fact [**EiM**], the pair (ω, γ) constitutes a 3-cocycle for a cohomology theory known as *abelian group cohomology*. Moreover, the map

$$q : H^3_{\mathrm{ab}}(A, \mathbb{K}^\times) \to \mathbb{K}^\times$$

that is given by $g \mapsto \gamma(g,g)$ is a *quadratic form* on the group A, i.e., obeys $q(g^{-1}) = q(g)$ and is such that the map $(g_1, g_2) \mapsto q(g_1g_2)\, q(g_1)^{-1}\, q(g_2)^{-1}$ is a group homomorphism in each argument. It follows that we can associate to a finite abelian group A and a quadratic form $q\colon A \to \mathbb{K}^\times$ a braided fusion category $\mathcal{C}(A, q)$ whose isomorphism classes of simple objects are in bijection with the elements of A. The braiding on $\mathcal{C}(A, q)$ is nondegenerate if and only if the quadratic form q is nondegenerate (see Example 8.13.5 of [**EtGNO**]). To obtain a ribbon category, additional data must be specified.

(2) Recall from Example 2.24 the notion of a Lie algebra over a field $\mathbb{K}$. A (nonzero) Lie algebra is called *simple* if it does not contain any ideal besides itself and the zero vector space. (For instance, the vector spaces $\mathfrak{sl}(n)$ of traceless $n \times n$-matrices, endowed with the commutator as a Lie bracket, are simple Lie algebras.) Consider a finite-dimensional Lie algebra $\mathfrak{g}$ over the complex numbers that is a finite direct sum $\mathfrak{g} = \bigoplus_i \mathfrak{g}_i$ of simple Lie algebras. To $\mathfrak{g}_i$ one associates an (infinite-dimensional) *untwisted affine* Lie algebra $\widehat{\mathfrak{g}}_i$ as a central extension of its loop algebra (see, e.g., [**Kac**] or Chapter 12 of [**FS1**]). An important subclass of representations of the Lie algebra $\widehat{\mathfrak{g}}_i$ are the integrable representations which, like the irreducible finite-dimensional representations of $\mathfrak{g}_i$ are *highest weight representations*. Such a representation can be characterized by a highest weight of $\mathfrak{g}_i$ and a number k_i, called the *level*, which for integrability must be a positive integer.

   For any collection $\{k_i\}$ of positive integers (one for each simple ideal $\mathfrak{g}_i$ of $\mathfrak{g}$) there exists a modular tensor category $\mathcal{C}(\mathfrak{g}, \{k_i\})$ whose simple objects are given by tuples of representations of $\widehat{\mathfrak{g}}_i$ that are integrable of level k_i. The modular data of $\mathcal{C}(\mathfrak{g}, \{k_i\})$ are, e.g., presented in Theorem 13.8 of [**Kac**]. Further details for the modular tensor category $\mathcal{C}(\mathfrak{sl}(2), k)$ can be found in Section 2.1 of [**FeFFS**] and Section 8.18.2 of [**EtGNO**].

(3) Also some more specific classes of modular fusion categories are known, such as [**ArCRW**] the so-called metaplectic modular categories.

(4) An important source for examples of nonsemisimple modular tensor categories is a construction that makes use of screening charges on vertex operator algebras; via Nichols algebras these are intimately related to representation categories of quantum groups [**Len**]. The modular tensor categories for so-called symplectic fermions can be regarded as special cases; see, e.g., [**Ru**]. A detailed treatment of such models goes beyond the level of the present notes.

We are now in a position to state the classification result for 3-2-1-dimensional topological field theories, which is stated in the unpublished preprint [**BDSV**]. The result does not literally apply to the type of 3-2-1-dimensional topological field theory given by Definition 5.18, but rather to a slight variant of it. More specifically, the groups of invertible three-dimensional cobordisms $\Sigma \longrightarrow \Sigma$ for every surface Σ – that is, the mapping class groups,, i.e., the groups of isotopy classes of self-diffeomorphisms of surfaces, see Remark 2.104 – must be enlarged by passing to a certain central extension. We refer to Chapter 4 of [**Tur1**] for details of this extension, to [**FaM**] for an exposition of the theory of mapping class groups, and to [**Fun**] for a review of mapping class group representations coming from topological field theories. By allowing for such a central extension we arrive at the notion of an *anomalous* 3-2-1-dimensional topological field theory. (The inclusion of the *anomaly* is standard, to the extent that the qualification 'anomalous' is frequently omitted even when, like for instance in the theories obtained by surgery constructions (see Section 6.4.3), strictly speaking it must be kept.) The field theories covered by Definition 5.18, without the anomaly, are then called *anomaly-free*. Our working example, Dijkgraaf-Witten theory, is anomaly-free, and accordingly we refrain from a further discussion of the anomaly here.

THEOREM 5.60 ([**BDSV**]). The evaluation of 3-2-1-dimensional topological field theories on the circle provides an equivalence

$$\left\{\begin{array}{c}\text{anomalous 3-2-1-dimensional}\\ \text{topological field theories}\\ \text{with values in 2-vector spaces}\end{array}\right\} \xrightarrow{\simeq} \left\{\begin{array}{c}\text{semisimple modular categories}\\ \text{in 2-vector spaces}\\ \text{with possibly nonsimple unit}\end{array}\right\}.$$

PROPOSITION 5.61. The topological field theory associated to a semisimple modular category $\mathcal{C}$ assigns to a compact oriented surface of genus g with p incoming and q outgoing boundary components the linear functor $\mathcal{C}^{\boxtimes p} \longrightarrow \mathcal{C}^{\boxtimes q}$ that sends the object $X_1 \boxtimes \cdots \boxtimes X_p$ to

$$\bigoplus_{i_1,\ldots,i_q=1}^{n} \operatorname{Hom}_{\mathcal{C}}(Y_{i_1} \otimes \cdots \otimes Y_{i_q}, X_1 \otimes \cdots \otimes X_p \otimes \mathbb{F}^{\otimes g})\, Y_{i_1} \boxtimes \cdots \boxtimes Y_{i_q}\,,$$

where $\mathbb{F} = \bigoplus_{i=1}^{n} X_i \otimes X_i^{\vee}$, and where the direct sum runs over a set of representatives for the isomorphism classes of simple objects of $\mathcal{C}$.

CHAPTER 6

More about TFTs: A few appetizers

Topological field theories, which have been studied intensively for more than three decades, are still under active investigation, both in mathematics and in physics. In this final chapter we outline a few directions of recent and ongoing research and make suggestions for further reading. Our main focus is on developments at the interface to category theory and representation theory. Needless to say, our selection is biased by the authors' taste and restricted knowledge.

Recall that a topological field theory is a symmetric monoidal functor

$$\mathcal{C}ob \longrightarrow \mathcal{V}$$

from some symmetric monoidal (possibly higher) category of cobordisms to a symmetric monoidal (higher) category $\mathcal{V}$. Various different cobordism categories will be involved in the subsequent considerations.

6.1. Modular functors

The notion of a 3-2-1-dimensional topological field theory has the following important weakening. In Theorem 5.55 we have seen that any 3-2-1-dimensional topological field theory leads, by evaluation on the circle, to a semisimple ribbon category. On the other hand, as has been known for a long time [**Lyu**], there are also nonsemisimple ribbon categories that allow for many constructions inherent in a topological field theory, including those described in Proposition 5.61. In particular one can still construct functors for oriented surfaces with boundary circles that carry an action of the mapping class group of surfaces, i.e., the group of orientation-preserving self-diffeomorphisms up to isotopy. Concretely, to any mapping class group element one may associate a natural isomorphism of functors in such a way that the composition of mapping classes is respected. Moreover, the so obtained linear functors associated to surfaces are also compatible with the gluing of surfaces; details can be found in [**Lyu**] and also, e.g., in Section 2 of [**FSS**] and in [**FScS**]. Together, the linear functors associated to surfaces, their mapping class group actions and their gluing properties form what is known as a *modular functor*; see, e.g., [**Til**, **BakK**]. A modular functor can be thought of as a 3-2-1-dimensional topological field theory that is not defined on all three-manifolds.

The mapping class group actions that arise in this way play a crucial role in various applications, such as for the design of logical gates in topological quantum computing (see, e.g., [**BlW**]) and for the construction of correlators in two-dimensional conformal field theories that was mentioned in Remark 3.13.

It is an obvious question to ask how a modular functor can be extended to a full-fledged 3-2-1-dimensional topological field theory. This question has been addressed, using a particular presentation of the oriented three-dimensional cobordism category, in the setting of string-net models in [**Bar**].

6.2. Boundaries and defects

In recent years it has become increasingly clear that manifolds with boundary and stratified manifolds should be admitted as the space on which a quantum field theory is formulated. Accordingly, in the case of topological field theories one also considers cobordism categories defined in terms of manifolds that can have boundaries or even corners and, moreover, come with distinguished submanifolds of various codimensions. The precise definition of a stratified manifold is subtle; we refrain from presenting it here and instead refer to Section 1 of [**AyFR**].

6.2.1. Two-dimensional topological field theory with boundaries.

For two-dimensional TFTs with boundaries, the appropriate symmetric monoidal cobordism category $\mathcal{C}ob^{\partial}_{2,1}$ has compact one-dimensional manifolds without boundary as well as compact one-dimensional manifolds with boundary as objects. An object of $\mathcal{C}ob^{\partial}_{2,1}$ is thus isomorphic to a finite disjoint union of copies of the standard interval $I = [0,1] \subset \mathbb{R}$ and of the standard circle $S^1 \subset \mathbb{C}$. A morphism $B\colon M \longrightarrow N$ between objects M and N consists of an oriented smooth surface B with boundary and corners and an orientation preserving embedding $\overline{M} \sqcup N \hookrightarrow \partial B$. Surfaces which are diffeomorphic relative to the parametrized boundary are considered as equivalent and describe the same morphism in $\mathcal{C}ob^{\partial}_{2,1}$.

The boundary ∂B of an object B is allowed to contain segments that are not parametrized. Those segments that *are* parametrized are called *gluing circles* respectively *gluing intervals*. Composition of morphisms in $\mathcal{C}ob^{\partial}_{2,1}$ is given by gluing along parametrized segments of the boundary, and the tensor product is disjoint union. We call the category $\mathcal{C}ob^{\partial}_{2,1}$ the category of *open-closed cobordisms*. $\mathcal{C}ob^{\partial}_{2,1}$ contains the category $\mathcal{C}ob_{2,1}$ discussed in Section 3.2 as a subcategory. For more details about $\mathcal{C}ob^{\partial}_{2,1}$, in particular a presentation in terms of generators and relations, we refer to [**LaP**].

A TFT on two-manifolds with boundary is defined as a symmetric monoidal functor

$$Z: \quad \mathcal{C}ob^{\partial}_{2,1} \longrightarrow \mathcal{V}ect(\mathbb{K}).$$

With the help of a presentation of the category $\mathcal{C}ob^{\partial}_{2,1}$, this functor can be studied in the spirit of the analysis that led to Theorem 3.14 relating two-dimensional topological field theories to commutative Frobenius algebras. Indeed, the evaluation of Z on S^1 gives again a commutative Frobenius algebra A, while the evaluation on the interval I gives a symmetric Frobenius algebra A_∂ that is not necessarily commutative. Additional structure comes from a cobordism called the *zipper*, that is, an annulus for which one boundary component is a gluing circle and the second boundary component contains one gluing interval. This provides two morphisms in $\mathcal{C}ob^{\partial}_{2,1}$, one for which the gluing circle is incoming and the gluing interval is outgoing, and one for which the interval is incoming and the circle is outgoing. Applying the functor Z to these morphisms gives two linear maps

$$\imath: \quad A \longrightarrow A_\partial \qquad \text{and} \qquad \bar\imath: \quad A_\partial \longrightarrow A,$$

which have to obey various consistency conditions. The resulting algebraic system is called a *knowledgeable Frobenius algebra* in [**LaP**]; we refer to that paper and to [**MooS1**] for further information.

6.2.2. Boundary conditions. A striking new feature of $\mathcal{C}ob^{\partial}_{2,1}$ is the presence of the unparametrized boundary segments, which are often called *physical boundaries*. In the setting of Section 6.2.1 the physical boundary segments of a cobordism are unlabeled. A further refinement is achieved by instead considering labeled physical boundaries, i.e., attaching to each physical boundary a value in some collection $\{a\}$ of labels; these labels are called *boundary conditions*. This raises immediately the question of how, based on the knowledge of the commutative Frobenius algebra assigned to the gluing circle S^1 or of the symmetric Frobenius algebra assigned to a gluing interval I, we can determine the set of all possible boundary conditions.

To address this issue we first notice that in the situation we are considering now, instead of having a single type of parametrized interval I, one effectively deals with a collection of gluing intervals I^a_b, with a and b the boundary conditions carried by the physical boundary segments adjacent to the endpoints of the gluing interval. By the same considerations as in the case of unlabeled physical boundaries, one then obtains a symmetric Frobenius algebra

$$A_a := Z(I^a_a)$$

for every boundary condition a, each of them endowed with maps $\bar{\imath}_a$ and $\imath_a$ to and from the bulk algebra $A = Z(S^1)$ as described in Section 6.2.1. In addition, we have vector spaces $M^a_b := Z(I^a_b)$ with $a \neq b$. Consider now a circle with three gluing intervals for which the three connected components of the complement of the gluing intervals are labeled by a, a and b, respectively. Taking the gluing intervals I^a_a and I^b_a to be incoming and the third interval I^b_a to be outgoing, we obtain an action of the algebra A_a on M^a_b, whereby M^a_b becomes a left A_a-module.

We thus get a linear category $\mathcal{C}$ whose objects are boundary conditions and whose morphism spaces are given by the vector spaces M^a_b. (Since the endomorphisms carry the additional structure of a Frobenius algebra, $\mathcal{C}$ has actually a bit more structure: it is a Calabi-Yau category.) In the case of the two-dimensional Dijkgraaf-Witten theory for a finite group G, the category $\mathcal{C}$ has been argued to be equivalent to the category G-$\mathcal{R}ep$ of finite-dimensional G-representations; more information, in particular concerning the relation with the bulk algebra. is given in Example 2.2 in [**MooS1**]. (For boundary conditions in three-dimensional Dijkgraaf-Witten theories see Section 6.2.5.)

6.2.3. Defect lines. Once we allow for the presence of (labeled) physical boundaries, it is quite natural to admit also "inner boundaries", or *defect lines*. We briefly note a few pertinent results: Such defect lines divide the interior of a cobordism into two-dimensional regions, to each of which we can assign a symmetric Frobenius algebra that lies in the Morita class of the boundary algebra A_a if the region is adjacent to a physical boundary with boundary condition a. In general, the Frobenius algebras on either side of a defect line need not belong to the same Morita class. While a boundary condition corresponds to a module over a boundary algebra that is determined up to Morita equivalence, a defect line is then naturally labeled by a bimodule over a pair of not necessarily Morita equivalent symmetric Frobenius algebras.

The investigation of defects of various codimensions in quantum field theories has in fact given rise to a shift of paradigm in the perception of symmetries. See, e.g., [**FS2**] for the case of three-dimensional topological field theories and [**Gom**] for developments in the study of more general quantum field theories.

6.2.4. Three-dimensional TFTs with boundaries. Boundaries can be considered for topological field theories of any dimension. A convenient starting point for an intuitive analysis of three-dimensional TFTs on manifolds with boundary is to determine the labels for various types of one-dimensional embedded submanifolds or, in physics terminology, of *Wilson lines*. One arrives at the following picture [**FSV1**].

For each boundary condition a there is a category $\mathcal{W}_a$ whose objects are the types of boundary Wilson lines that are located in the interior of a boundary component labeled by the boundary condition a. Since bulk and boundary are oriented manifolds, the category $\mathcal{W}_a$ should be a spherical fusion category. As the boundary Wilson lines are confined to a two-dimensional manifold, $\mathcal{W}_a$ does not come with the additional structure of a braiding.

The corresponding category $\mathcal{C}$ of bulk Wilson lines is naturally a braided category. Postulating the existence of a "smooth" process of moving a bulk Wilson line to the boundary, for every boundary condition a one obtains a functor

$$F_a : \quad \mathcal{C} \longrightarrow \mathcal{W}_a .$$

Imposing on the functor F_a a maximal number of naturality properties singles out a distinguished class of boundary conditions. This leads to the conclusion [**FSV1**] that $\mathcal{C}$ should be braided equivalent to the Drinfeld center $\mathcal{Z}(\mathcal{W}_a)$ of $\mathcal{W}_a$ and F_a the functor that forgets the half-braiding.

The category $\mathcal{W}_a$ is naturally a module category over $\mathcal{C}$. However, notably, not every $\mathcal{C}$-module category can be obtained this way. For instance, when $\mathcal{W}_a$ is the category of finite-dimensional representations of the finite group $\mathbb{Z}_2$, so that $\mathcal{Z}(\mathcal{A})$ is the category of finite-dimensional representations of the Drinfeld double $\mathcal{D}(\mathbb{Z}_2)$, then there are two indecomposable $\mathcal{W}_a$-module categories, but six indecomposable $\mathbb{Z}(\mathcal{W}_a)$-module categories (see [**Ost**, Thm. 3.1]).

The following new phenomenon arises. While every commutative algebra is the center of some associative algebra (namely of itself), a braided category, even if its braiding is nondegenerate, is not necessarily equivalent to the Drinfeld center of any monoidal category. In fact, nondegenerately braided finite tensor categories admit an equivalence relation such that the set of equivalence classes has a group structure, with the class of Drinfeld centers as its neutral element. This *Witt group* of nondegenerately braided tensor categories is a highly interesting object; for details we refer to [**DaMNO**]. The nontriviality of the Witt group implies that there exist three-dimensional topological field theories that do not admit any boundary condition of the type we described (see, e.g., [**FSV1, FrT**] for details). In short, the Witt class is an obstruction to the existence of a consistent topological boundary condition for a three-dimensional topological field theory.

Once one fixes a boundary condition by specifying a spherical fusion category $\mathcal{W}_{a_\circ}$ that satisfies $\mathcal{C} \simeq \mathcal{Z}(\mathcal{W}_{a_\circ})$, every other boundary condition is an object in the bicategory of spherical $\mathcal{W}_{a_\circ}$-module categories. In this manner, three-dimensional topological field theory furnishes a way to organize the theory of (bi)module categories over a spherical monoidal category. The latter system of categories can be understood as a semisimple 2-category [**DoR**].

Recently, activity also turned to the study of four-dimensional topological field theories, for which both the geometric and the categorical dimension are further increased. The input datum is then a *fusion 2-category*, and the theory provides insight into the structure formed by representation categories of braided tensor categories [**JR**, **Dec**].

6.2.5. Dijkgraaf-Witten theories. Next we indicate how the presence of boundaries and defects is accounted for in the particular case of our central example, i.e., for three-dimensional topological field theories of Dijkgraaf-Witten type.

Geometrically, boundaries and defects are particular instances of submanifolds. Accordingly they are captured by the notion of a *relative* (smooth) *manifold* $Y \xrightarrow{j} X$, which by definition is a pair Y, X of smooth manifolds together with a morphism $j\colon Y \longrightarrow X$ of smooth manifolds. In a Dijkgraaf-Witten theory on a manifold X with boundary, the pertinent data are a relative manifold $\partial X \xrightarrow{j} X$ and a morphism $H \xrightarrow{\varphi} G$ of finite groups. The field configurations then form the groupoid of *relative bundles*. The objects of this groupoid are triples consisting of a G-bundle $P_G \longrightarrow X$, an H-bundle $P_H \longrightarrow \partial X$ and an isomorphism

$$\alpha: \quad \mathrm{Ind}_H^G(P_H) \xrightarrow{\cong} j^* P_G$$

of G-bundles on ∂X, where the functor $\mathrm{Ind}_H^G\colon \mathcal{B}un_H(X) \longrightarrow \mathcal{B}un_G(X)$, which acts on objects as $P_H \longmapsto P_H \times_\varphi G$, is the extension functor considered in Exercise 1.46.

A morphism in the groupoid is a pair consisting of a morphism $P_G \longrightarrow P'_G$ of G-bundles on X and a morphism $P_H \longrightarrow P'_H$ of H-bundles on ∂X satisfying an appropriate compatibility constraint. Additional data from groupoid cohomology can enter the linearization process, which otherwise follows the same procedure as in the case of manifolds without boundary. In the framework of Lagrangian field theory, these cohomological data provide a topological bulk Lagrangian together with compatible boundary terms.

This analysis can be made fully explicit and can be extended to three-manifolds with embedded surface defects. For details, including in particular the explicit form of the groupoids obtained in specific situations, we refer to [**FSV2**] and Section 3.4 of [**FS2**]. We finally mention that boundary conditions and surface defects have also been studied in other settings. For instance, a heuristic study using path integrals in the context of (abelian) Chern-Simons theories, which we mentioned in Remark 2.2, is presented in [**KapS**].

6.3. Tangential structures

So far we have been discussing topological field theories that are based on *oriented* manifolds. We have, e.g., seen that two-dimensional oriented topological field theories correspond to Frobenius algebras.

From a geometric perspective this raises the following issues: Manifolds do not necessarily have an orientation; see, e.g., exercise 1.21. In particular, cobordism categories can also be constructed by considering spans of unoriented manifolds. Manifolds can also have additional structure besides an orientation, for instance a spin structure, and again corresponding categories of cobordisms can be defined. Accordingly one should expect that different flavors of topological field theories exist that correspond to different tangential structures of the underlying manifolds. One then wants to understand the relation between different flavors of topological field theories, as well as the corresponding algebraic structures. Conversely, since Frobenius algebras are algebras with additional structure, it is an obvious question to ask which tangential structures lead to topological field theories whose evaluation on circles yields just algebras rather than Frobenius algebras.

6.3.1. Unoriented topological field theories. As in the oriented case, the analysis of an unoriented two-dimensional topological field theory can start with a combinatorial description of the (bi)category of (extended) cobordisms in terms of generators and relations. We refer to Section 3.8.6 of [**Scho**] for a detailed analysis, which in particular shows that fully extended unoriented two-dimensional topological field theories correspond to *stellar separable Frobenius algebras.* A stellar algebra is an algebra A equipped with a Morita context s between A and its opposite algebra A^{opp}, as well as with an isomorphism of Morita contexts between s and another Morita context between A and A^{opp}, called the opposite Morita context, subject to certain consistency conditions; for details see Definition 3.78 of [**Scho**].

6.3.2. Framed topological field theories. To obtain two-dimensional topological field theories whose evaluation on the circle has just the structure of a unital associative algebra, and thus *less algebraic* structure than the one of a Frobenius algebra, one must consider two-manifolds with *more geometric* structure, namely with a 2-framing. As has been mentioned in Remark 1.40, a 2-framed two-manifold is a two-manifold M that is endowed with two nonvanishing vector fields that form a basis of the tangent space in each point of M. To get a feeling for the cobordism category $\mathcal{C}ob_{2,1}^{\mathrm{fr}}$ of 2-framed manifolds, it is helpful to imagine the one-manifolds S that constitute the objects of $\mathcal{C}ob_{2,1}^{\mathrm{fr}}$ to come with a two-dimensional collar, so that the notion of two linearly independent vector fields makes sense. Alternatively one can consider a trivialization of the direct sum of the tangent bundle $\mathrm{T}S$ and the trivial bundle $\underline{\mathbb{R}}$ on S, i.e., an isomorphism

$$\mathrm{T}S \oplus \underline{\mathbb{R}} \cong \underline{\mathbb{R}}^2.$$

As a consequence of these additional geometric data, there is now a $\mathbb{Z}$-family of 2-framed circles $S^1_{(n)}$, where $n \in \mathbb{Z}$ is a winding number for the two nonzero vector fields. The existence of a 2-framed pair of pants with such circles as boundary components is highly restricted. Therefore only one of the circles $S^1_{(n)}$ gives rise to an algebra A which, however, does not come with the structure of a Frobenius algebra. The other framed circles are mapped to A-bimodules.

Framed topological field theories turn out to be central for the understanding of the structure of all types of topological field theories. In particular, one of the most fundamental results in topological field theory, the *cobordism hypothesis*, is most directly stated for framed (fully extended) topological field theories. The cobordism hypothesis is first of all a structural result about framed n-dimensional manifolds, which we now briefly explain.

Recall the combinatorial description of the symmetric monoidal category $\mathcal{C}ob_{2,1}$ in Section 3.2. That result can be reformulated as the statement that $\mathcal{C}ob_{2,1}$ is the free symmetric monoidal category containing a (co)commutative Frobenius object, see Theorem 3.6.19 of [**Koc**]. This statement, in turn, readily implies that two-dimensional topological field theories with values in a symmetric monoidal category

$\mathcal{V}$ are classified by the groupoid of commutative Frobenius algebras in $\mathcal{V}$. In a similar vein, the cobordism hypothesis states that the (∞, n)-category of framed n-dimensional cobordisms is equivalent to the free symmetric monoidal (∞, n)-category with duals on a single object.

Here it is indeed necessary to use the framework of ∞-categories. This is beyond the scope of these lecture notes, but it is worth pointing out that ∞-categories arise quite naturally. Indeed, recall from Remark 2.9 that an extension of the cutting process of manifolds that allows for codimension-2 submanifolds leads to bicategories. A further extension involving submanifolds of codimension higher than 2 requires still higher categories. These are, at least according to our present understanding, best dealt with as ∞-categories.

The extreme case of an extension is down to a zero-dimensional submanifold, i.e., a point. In this case the structural insight into framed n-dimensional manifolds entails that the evaluation functor

$$Z \longmapsto Z(*)$$

furnishes a bijection between framed fully extended $\mathcal{V}$-valued topological field theories and isomorphism classes of fully dualizable objects of $\mathcal{V}$. A dualizable object in a symmetric monoidal (∞, n)-category $\mathcal{V}$ is called *fully dualizable* if the structure maps of the duality unit and counit themselves have adjoints, the unit and counit of which, in turn, have adjoints as well, and so on. For more details we refer the reader to [**Lur**] and [**Fr2**].

Based on this understanding of framed topological field theories, one can study topological field theories with additional tangential structures, e.g., an orientation or a spin structure. These structures can be realized as fixed points of group actions. In the case of two-dimensional topological field theories, this leads to the insight that finite-dimensional semisimple symmetric Frobenius algebras can be described as homotopy fixed points of an SO(2)-action on the bicategory of finite-dimensional semisimple algebras, bimodules and intertwiners; for details see, e.g., [**HeSV**].

6.3.3. Equivariant topological field theories. From the point of view of representation theory, the following tangential structure is particularly interesting: Fix a finite group G and endow every manifold with the additional structure of a G-bundle. This yields a cobordism category $\mathcal{C}ob^G_{n,n-1}$. We restrict our attention to the case of a three-dimensional (once-)extended TFT in the sense of Definition 5.18. Then to a circle S^1 we need to assign a category together with a G-bundle. Characterizing such a G-bundle through its holonomy and thus by a group element $g \in G$ (and trivializing the fiber over a single point), we obtain a category $\mathcal{C}_g$ for each group element $g \in G$. The collection of these categories is endowed with further structure: G-bundles on pairs of pants provide tensor product functors

$$\mathcal{C}_g \times \mathcal{C}_h \longrightarrow \mathcal{C}_{gh}$$

with appropriate coherence data, for all $g, h \in G$. (Note that G-bundles over the three boundary components determine a G-bundle on a pair of pants up to isomorphism, if such a bundle exists.) As a consequence, each of the categories $\mathcal{C}_g$ is a $\mathcal{C}_e$-bimodule category, with e the unit element of G. A second set of data is provided by cylinders over S^1 with a G-bundle; they give coherent G-actions

$$\rho_h : \quad \mathcal{C}_g \longrightarrow \mathcal{C}_{hgh^{-1}} .$$

In this way, the category $\bigoplus_{g\in G}\mathcal{C}_g$ is endowed with the structure of a *G-crossed category* with a *G-braiding.* For more information on this type of topological field theories we refer to [**Tur2**].

Conversely, as explained, e.g., in [**EtNOM**], a G-crossed modular tensor category can be constructed from a suitable collection of $\mathcal{C}_e$-bimodule categories together with a few cohomological data. In case the modular category admits a concrete representation-theoretic realization as a category of representations over a rational vertex algebra, then such systems can be obtained from representations of that vertex algebra that are twisted by an automorphism of the vertex algebra [**McR**].

6.3.4. Orbifolds / equivariantization. In Examples 5.56 and 5.59 we have presented several classes of (semisimple) modular tensor categories. As that list indicates, the supply of algebraic structures with representation categories that come with a natural structure of a modular tensor category is, at least for the moment, quite limited. Accordingly it is of much interest to find constructions of new modular tensor categories out of known ones. Here we explain a procedure that produces from a G-modular tensor category $\widetilde{\mathcal{C}}$ a modular tensor category $\mathcal{C}=\widetilde{\mathcal{C}}//G$. This construction is called *equivariantization* in mathematics and *orbifold construction* in the physics literature. The objects of the category $\widetilde{\mathcal{C}}//G$ are fixed points of the G-action on the objects of $\widetilde{\mathcal{C}}$. According to the principle, formulated in Remark 1.12, that it is generally inappropriate to regard two objects as equal rather than as isomorphic, a fixed point $X\in\widetilde{\mathcal{C}}$ comes with extra data, namely with isomorphisms

$$\alpha_h:\quad \rho_h(X)\xrightarrow{\cong}X$$

for all $h\in G$, obeying a suitable compatibility condition.

There is a procedure that is inverse to equivariantization: The category $\widetilde{\mathcal{C}}//G$ contains a full symmetric monoidal subcategory that is equivalent to G-$\mathcal{R}ep$, which arises from equivariantizations on direct sums of the monoidal unit of $\widetilde{\mathcal{C}}$. This subcategory contains a canonical commutative algebra object A_G, induced from the algebra of functions on G. The category $\widetilde{\mathcal{C}}$ can be reconstructed by considering A_G-modules in $\widetilde{\mathcal{C}}//G$. For details we refer to [**GelNN**].

6.4. Topological field theories from algebraic data

In this section we outline several constructions of topological field theories from algebraic input data. The focus will be on oriented three-dimensional topological field theories.

6.4.1. String-net models. We first describe the *string-net* construction [**LeW**, **Kir**, **Bar**] of oriented three-dimensional TFTs, which takes a spherical fusion category as algebraic input datum. Similar constructions are expected to work in any number d of dimensions; in the four-dimensional case, a ribbon fusion category is the appropriate input datum.

The first step in a string-net construction consists of assigning vector spaces to $(d-1)$-dimensional manifolds. In the sequel we restrict ourselves to the case $d=3$, so that we have to assign vector spaces to oriented surfaces Σ (which may have a boundary and may also be noncompact). To this end one starts with the vector space $\mathbb{K}\mathrm{Gr}(\Sigma)$ that is freely generated on the set of finite $\mathcal{C}$-labeled graphs Γ on Σ. If Σ has nonempty boundary, then on each connected component of $\partial\Sigma$ at least one dangling edge has to end, and the labels of all edges ending on $\partial\Sigma$ have to be kept fixed.

To proceed, a crucial tool is an appropriate graphical calculus. For any pivotal tensor category $\mathcal{C}$ a graphical calculus can be formulated on an oriented canvas and thus in particular on disks [**Kir**]. (An analogous graphical calculus can be set up for the more general input datum of a pivotal bicategory [**FSY**].) This calculus can be used to replace any finite graph on a disk D having edges labeled by objects of $\mathcal{C}$ and vertices labeled by compatible morphisms of $\mathcal{C}$ and having dangling edges on ∂D by a star-shaped graph on D. Such a graph amounts, in turn, to a morphism in $\mathcal{C}$, whereby the prescription canonically defines a surjection $\langle - \rangle_D$ from the vector space $\mathbb{K}\mathrm{Gr}(D)$ freely generated by graphs on the disk to the space of morphisms in $\mathcal{C}$ having as domain the monoidal unit $\mathbf{1}$ and as codomain the cyclic tensor product of the objects labeling the dangling edges on ∂D.

For any surface Σ one now considers the subspace $\mathrm{N}(\Sigma) \subset \mathbb{K}\mathrm{Gr}(\Sigma)$ that consists of all *null graphs*, i.e., of all linear combinations $\sum_i \lambda_i \Gamma_i$ of graphs that coincide outside any chosen embedded disk $D \subset \Sigma$ and whose restrictions to D sum up to zero, in the sense that

$$\sum_i \lambda_i \langle \Gamma_i \cap D \rangle_D = 0 .$$

The vector space assigned to Σ is then the quotient

$$\mathrm{SN}(\Sigma) := \mathbb{K}\mathrm{Gr}(\Sigma) \,/\, \mathrm{N}(\Sigma)$$

of $\mathbb{K}\mathrm{Gr}(\Sigma)$ by the subspace of null graphs, which turns out to be finite-dimensional.

There exist different flavors of graphical calculus, and correspondingly different variants of string-net models. Pivotal categories lead, as already stated, to a graphical calculus on an oriented canvas and thus to modular functors on oriented manifolds. Waiving the pivotal structure requires to assume more geometric structure. Indeed, then the appropriate two-dimensional graphical calculus is in terms of progressive graphs [**JS1**], which are naturally drawn on a 2-framed canvas; corresponding string-net models on 2-framed two-manifolds have been studied in [**KnST**].

The $(d-1)$-dimensional manifold Σ entering in the string-net construction can have boundaries and even corners. By considering such manifolds one can extract also the algebraic data that are assigned to lower-dimensional manifolds. For instance, for $d=3$ the category assigned to a circle S^1 is extracted as follows: An object is a collection of finitely many points on the circle, each labeled by an object in $\mathcal{C}$; a morphism is a string net on the cylinder over the circle with appropriate labels on the two boundary circles. This category can be identified with a full monoidal subcategory of the Drinfeld center $\mathcal{Z}(\mathcal{C})$ of $\mathcal{C}$. For $d=3$, an extension to three-dimensional manifolds, and thus from a modular functor to a three-dimensional TFT is possible as well, provided that the input category is a semisimple spherical category; for details we refer to [**Bar**].

The string-net construction is an example of a skein-theoretic construction. Skein-theoretic methods frequently appear in topological field theory. In fact, the HOMFLY knot polynomial, which generalizes the Jones polynomial, is defined by such methods. Skein-theoretic methods also allow one to integrate braided categories (with some additional properties) over framed two-dimensional manifolds [**BeBJ**]. This can be seen as a realization of factorization homology; the relation between factorization homology and topological field theory is explained in [**Sche**].

6.4.2. State-sum models. String-net models admit various reformulations; in particular they can be presented in the form of state-sum models. Such models [**TV1**] are historically among the first topological field theories [**TV2**, **BarW**], with precursors in work on three-dimensional quantum gravity [**PR**]; they are also called *Turaev-Viro TFTs.* In a state-sum model the invariant assigned to a closed oriented three-manifold M is constructed using the auxiliary datum of a *skeleton* in M. A skeleton is an oriented stratified 2-polyhedron embedded in M which provides a combinatorial model for M; a skeleton can be thought of as a system of submanifolds of various dimensions, of the type illustrated in the following picture:

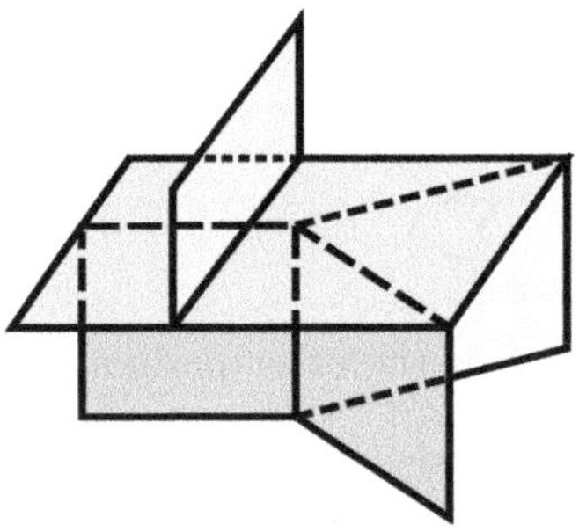

A detailed definition of a skeleton is given in Chapter 11 of [**TV1**]; a reader familiar with basic notions of algebraic topology may imagine, for simplicity, the structure of a finite CW complex on M. The term *state-sum* model originates in the fact that one considers all *colorings* of the skeleton Δ, i.e., all assignments c of isomorphism classes of simple objects of $\mathcal{C}$ to the 2-cells of Δ. By evaluating the graphs on spheres that surround the 0-cells of Δ, each coloring c yields a number $|c|$. One then defines for any spherical category $\mathcal{C}$ and closed oriented three-manifold M with skeleton Δ the sum

$$|M|_{\mathcal{C}} := (\dim \mathcal{C})^{-|M\setminus\Delta|} \sum_{c} \dim(c)\, |c|\,,$$

where $\dim(c)$ is the product over the dimensions of the simple objects $c(r)$ coloring the 2-cells r of Δ, raised to the power of the Euler characteristic of r, $\dim \mathcal{C}$ is the sum of squared dimensions of a representative set of simple objects of $\mathcal{C}$, and $|M\setminus\Delta|$ is the number of connected components of $M\setminus\Delta$. The number $|M|_{\mathcal{C}}$ can be shown to be independent of the choice of skeleton Δ (see Theorem 13.1 of [**TV1**]). It is the invariant assigned by the state-sum topological field theory based on the spherical fusion category $\mathcal{C}$ to the three-manifold M.

The vector space associated to a surface Σ is constructed as follows. Given a choice of a two-dimensional skeleton Δ_2 on Σ, one first constructs from the morphism spaces of $\mathcal{C}$ an auxiliary vector space $V(\Sigma, \Delta_2)$ which depends on Δ_2.

Then one uses cylinders over Σ to obtain an idempotent p on $V(\Sigma, \Delta_2)$. The image

$$V(\Sigma) := \mathrm{Im}\,(p) \subset V(\Sigma, \Delta_2)$$

of that idempotent can be shown to be independent of the choice of skeleton Δ_2; it is the vector space assigned to the surface Σ by the state-sum topological field theory.

State-sum constructions are also connected with *tensor networks*. Tensor network methods are an important tool in the study of strongly correlated systems in physics. They allow one, e.g., to circumvent the problem of exponential growth of the size of the state space when constructing ground states for quantum mechanical many-particle systems. For an introduction to such methods see, e.g., [**Oru, BrC, CiPSV**]. Two-dimensional tensor networks can be understood in terms of three-dimensional state-sum models on manifolds with physical boundaries. As shown in Section 5 of [**LoFHSV**], in this picture Wilson lines in the physical boundaries encode so-called MPO-symmetries of the space of ground states of the tensor network.

Finally it is worth pointing out that working with isomorphism classes of simple objects and summing over them is not well in line with the spirit of categories. This observation is relevant for string-net constructions based on nonsemisimple input data. In that case the construction cannot be expected to extend to all top-dimensional manifolds. In such constructions the sum over isomorphism classes of simple objects is replaced [**FScS**] by the categorical notion of a *coend*.

6.4.3. Surgery. It is natural to ask whether there are three-dimensional topological field theories that can be constructed by taking as an input a modular tensor category that is not necessarily a Drinfeld center. Such TFTs indeed exist, provided that one works with a slightly weakened variant of the TFT functor, namely with an *anomalous* TFT, in which – as mentioned in the paragraph preceding Theorem 5.60 – one allows for a projective assignment of data. Specifically, the representation of the mapping class group $\mathrm{Map}(\Sigma)$ of a surface Σ that is induced by cylinders over Σ is then only a projective representation.

Such a *Reshetikhin-Turaev* TFT, first developed [**ReT**] for input categories coming from quantum groups, can be constructed from any modular fusion category. In the case of modular fusion categories obtained from compact Lie groups, this leads to an algebraic construction of Chern-Simons topological field theories. For obtaining invariants of three-manifolds, the Reshetikhin-Turaev construction uses surgery along links. Indeed, any closed three-manifold can be obtained from the three-sphere S^3 by removing a tubular neighborhood of a framed link and gluing that neighborhood back by using an element of the mapping class group of the surface that is encoded in the link. For details on the construction we refer to [**Tur1**]. Surgery and state state sum constructions for three-dimensional topological field theories are reviewed in [**SöV**]. A modern point of view is to see a Reshetikhin-Turaev TFT as the boundary theory of a four-dimensional invertible topological field theory, the Crane-Yetter theory [**BarFG**].

Three-manifold invariants based on nonsemisimple modular tensor categories have been constructed in [**CGP, BCGP**] with the help of the so-called *modified trace* [**GeePT**]; for a partial extension of these invariants to a three-dimensional topological field theory see [**DGGPR**].

6.4.4. Topological field theories of cohomological type. We finally mention a construction of topological field theories that has its roots in the study of supersymmetric quantum field theories. From a physics perspective, this type of theories can be motivated as follows. A common feature of quantum field theories is the presence of a *stress-energy operator*; one crucial property of this field is that taking some kind of Lie bracket with the fundamental fields of the theory amounts to acting with a translation operator. A simple way to obtain a topological field theory is to demand the stress-energy operator to vanish. However, in situations in which the fields are not organized in terms of vector spaces, but in terms of complexes of vector spaces, a weaker requirement is sufficient, namely that the stress-energy operator, while nonzero, is *exact*. This condition can in particular be satisfied, via what is called a *topological twist*, in quantum field theories with extended supersymmetry. For such theories the generators of the supersymmetry algebra can be used to express a modification of the original stress-energy tensor as a derivative and thus as an exact operator. Closed operators then furnish exactly solvable sectors of the theory. Correspondingly, such topological field theories have been termed *cohomological* (topological) field theories.

Making these physics ideas mathematically precise leads to the notion of a *topological conformal field theory*. As formalized long ago in the *Segal axioms* (see [**Seg**] for the most complete available version), a two-dimensional conformal field theory should be a symmetric monoidal functor with domain a category $\mathcal{C}ob_{2,1}^{\mathrm{conf}}$ that has the same objects – disjoint unions of circles and intervals – as the domain $\mathcal{C}ob_{2,1}$ of a two-dimensional topological field theory (as treated in Section 3.2), but whose morphisms are Riemann surfaces, rather than cobordisms, having these objects as incoming and outgoing boundaries.

By taking instead as morphisms *moduli spaces* of Riemann surfaces, $\mathcal{C}ob_{2,1}^{\mathrm{conf}}$ is promoted to a cobordism category $\mathcal{C}ob_{2,1}^{\mathrm{conf\text{-}top}}$ that is enriched over the category $\mathcal{T}op$ of topological spaces. An efficient categorical language for addressing the so obtained structures is provided by $(\infty,1)$-categories. In this setting, a topological conformal field theory would then be something like an $(\infty,1)$-functor from $\mathcal{C}ob_{2,1}^{\mathrm{conf\text{-}top}}$ to some other infinity-category. However, yet another step (also anticipated in [**Seg**]) can be taken, namely a linearization. In modern terms, this consists in replacing $\mathcal{C}ob_{2,1}^{\mathrm{conf\text{-}top}}$ by a stable $(\infty,1)$-category $\mathcal{C}ob_{2,1}^{\mathrm{conf,dg}}$ whose morphisms are the homology chain complexes of those of $\mathcal{C}ob_{2,1}^{\mathrm{conf\text{-}top}}$. Finally, then, a topological conformal field theory is a symmetric monoidal $(\infty,1)$-functor from $\mathcal{C}ob_{2,1}^{\mathrm{conf,dg}}$ to the $(\infty,1)$-category of complexes of vector spaces.

For an exposition of cohomological field theories we refer to [**Co**] and [**LaL**].

6.5. Some applications

Applications of topological field theories, both in mathematics and in physics, abound. Here we present a limited and highly biased selection.

One important use of topological field theories is to provide a laboratory for exploring general structures and concepts in quantum field theories. An issue that we have stressed on various occasions is the notion of locality. A more recent illustration is provided by a mathematically rigorous approach to dimensional reduction [**MüW**], a procedure that assigns to an (extended) topological field theory of dimension d a topological field theory of dimension $d' < d$. Another example, which has experienced a surge of interest in recent years, are generalized symmetries; see, e.g., [**Scha, BhBF+**] for reviews. Such symmetries have made their first appearance in two-dimensional rational conformal field theories [**FroFRS**]. For establishing them in that context, substantial use is made of the close connection between three-dimensional topological field theories of surgery type and chiral rational conformal field theories. This relationship has played an important role since the early days of the subject [**Wit, MooS2**] and continues to be instrumental for the understanding of correlators in local conformal field theories (compare Remarks 2.39 and 3.13).

In addition, it finds a direct physical incarnation in quantum Hall systems in condensed matter physics. These are (quasi-)two-dimensional systems subject to a strong magnetic field. It can be argued [**FrohK**] that in the scaling limit a quantum Hall system, together with its time evolution, is captured by a Chern-Simons theory or some similar three-dimensional topological field theory. The discreteness of the values of the Hall conductivity in a quantum Hall system is in this approach a consequence of the discreteness of quantities like twist eigenvalues in the relevant modular fusion category. The presence of a magnetic field singles out an orientation in the two-dimensional sample, which is physically reflected by the orientation of the edge currents that exist in the sample. Accordingly, *chiral* rather than local conformal field theories are relevant for the description of universality classes of quantum Hall liquids.

A different type of applications is obtained in the context of state-sum constructions. As we have seen, such a construction assigns to a surface Σ and a skeleton Δ in Σ a vector space $V(\Sigma,\Delta)$ and a subspace $V(\Sigma)\subset V(\Sigma,\Delta)$ that does not depend on Δ. This is exactly the setup needed for formulating a *quantum code*. In that setting, the dimension of the subspace $V(\Sigma)$ is the number of qubits of the code. This point of view has, e.g., triggered the development of gauge theoretic models based on a semisimple finite-dimensional unimodular Hopf algebra H, known as *Kitaev models*. In these, the subspace is characterized as the space of ground states for a family of commuting Hamiltonians that are obtained with the help of a nonzero integral and cointegral of H, i.e., elements $\Lambda\in H$ and $\lambda\in H^*$ satisfying $\mu\circ(\mathrm{id}\otimes\Lambda)=\Lambda\circ\varepsilon$ and $(\lambda\otimes\mathrm{id})\circ\Delta=\eta\circ\lambda$, respectively. For details about such models we refer to [**Kit, BuMCA**].

The quasi-particle excitations in a Kitaev model (and likewise, the quasi-particle excitations in quantum Hall liquids) obey *braid group statistics*. In the physics literature, quasi-particles carrying a one-dimensional representation of the braid group are known as *anyons*, while those carrying a higher-dimensional representation are called non-abelian anyons. (The term *anyon* indicates that the spin, which parametrizes the phase shift of quasi-particle wave functions upon permutation, can take any value.) The same phenomenon appears in a much wider class of (effectively) two-dimensional quantum systems. The types of quasi-particles in such systems are expected to be classified as (isomorphism classes of) objects of a modular fusion category. In fact, a modular fusion category can be regarded as an invariant associated to a topological quantum phase of matter. This relationship is still under active study. For a guide to some recent developments in this area we refer to, e.g., [**ChTSR, Wen, Oga, McG**].

Topological field theories also provide a natural point of view on structures appearing in representation theory. One such structure is given by equivariant Frobenius-Schur indicators. The classical *Frobenius-Schur indicator* assigns to a $\mathbb{C}$-valued irreducible representation V of a finite group G the number

$$\mathrm{FS}(V) := \frac{1}{|G|} \sum_{g\in G} \chi_V(g^2)\,,$$

where χ_V is the character of V. The number $\mathrm{FS}(V)$ turns out to take one of the three values $+1$, 0, or -1, indicating whether the representation V is real, complex or quaternionic. The definition of $\mathrm{FS}(V)$ can be extended to simple objects in any pivotal tensor category $\mathcal{C}$, such as the category of finite-dimensional representations of $\mathrm{SU}(2)$, for which the simple objects are real and quaternionic, respectively, depending on whether their spin is integral or half-integral (compare Remark 1.12).

This generalizes further to the notion of an *equivariant* Frobenius-Schur indicator which assigns a number $\mathrm{FS}^{[Z]}_{(n,r)}(V)$ to an object $V\in\mathcal{C}$, an (isomorphism class of an) object Z in the Drinfeld center $\mathcal{Z}(\mathcal{C})$ of $\mathcal{C}$, and a pair of integers n and k [**NgS2**]. Now the modular group $\mathrm{SL}(2,\mathbb{Z})$ naturally acts both on the isomorphism classes of objects of the modular tensor category $\mathcal{Z}(\mathcal{C})$ and, through $2{\times}2$-matrices, on the pair (n,k). These two actions are intertwined by the equivariance relation

$$\mathrm{FS}^{\gamma.[Z]}_{(n,r)}(V) = \mathrm{FS}^{[Z]}_{(n,r).\gamma}(V)$$

for $\gamma\in\mathrm{SL}(2,\mathbb{Z})$. This makes the Frobenius-Schur indicator computable even for very big finite groups. This equivariance property has been explained with the help of the Turaev-Viro TFT based on the pivotal category $\mathcal{C}$, evaluated on a solid torus with physical boundary and boundary Wilson lines [**FaS**].

Another mathematical application is concerned with a natural notion of an embedding of a braided tensor category $\mathcal{B}$ into a nondegenerately braided tensor category $\mathcal{M}$ being *minimal*. Note that $\mathcal{B}$ embeds into its Drinfeld center $\mathcal{Z}(\mathcal{B})$ via the braiding, but this embedding is usually far from minimal. The existence of minimal extensions is important for the understanding of braided categories because tools for studying nondegenerately braided categories are far better developed than those for generic braided categories. The existence problem was solved [**JR**] in a surprising way by relating it to the fusion 2-category $\mathcal{M}od$-$\mathcal{B}$ of module categories over $\mathcal{B}$: minimal nondegenerate extensions of $\mathcal{B}$ correspond to certain trivializations of the Drinfeld center $\mathcal{Z}(\mathcal{M}od\text{-}\mathcal{B})$ which is a braided fusion 2-category whose objects label bulk surface operators in a (putative) four-dimensional topological field theory. The evaluation of this four-dimensional topological field theory on Klein bottles was used in [**JR**] to show that every slightly degenerate braided fusion category admits a minimal nondegenerate extension.

As a final example illustrating the explanatory power of topological field theory we mention *Radford's* s^4*-Theorem*. In its classical form this theorem states that for any finite-dimensional Hopf algebra H over a field $\mathbb{K}$ there are distinguished elements $a\in H$ and $\alpha\in H^*$ such that the identity

$$\mathrm{s}^4(h) = a\,(\alpha^{-1}\rightharpoonup h \leftharpoonup \alpha)\,a^{-1} = \alpha^{-1}\rightharpoonup (a\,h\,a^{-1}) \leftharpoonup \alpha$$

for the fourth power of the antipode s holds for every $h\in H$. Here $\rightharpoonup$ and $\leftharpoonup$ are right and left actions of the dual H^* on H; for details see, e.g., Section 7.1 of [**DaNR**]. This result can be generalized to the statement that for any finite tensor category $\mathcal{C}$ – in the classical case, the category of finite-dimensional H-modules – there is a distinguished isomorphism r of tensor functors with components

$$r_X : \quad X \xrightarrow{\cong} D \otimes {}^{\vee\vee\vee\vee}X \otimes D^{-1}$$

for $X \in \mathcal{C}$, where $D \in \mathcal{C}$ is an invertible object canonically associated with $\mathcal{C}$; see Section 7.19 of [**EtGNO**].

This purely algebraic fact can be understood in terms of the cobordism hypothesis for three-dimensional topological field theory, which we mentioned in Section 6.3.2. The orthogonal group O(3) acts on the cobordism category $\mathcal{C}ob^{\mathrm{fr}}_{\infty,3}$ by a change of the framing, and thus also acts on the functor category $\mathcal{F}un(\mathcal{C}ob^{\mathrm{fr}}_{\infty,3}, \mathcal{V})$, where the codomain $\mathcal{V}$ is taken to be the symmetric monoidal 3-category $\mathcal{B}imod$ that has finite tensor categories as objects, bimodule categories over finite tensor categories as 1-morphisms, and bimodule functors and bimodule natural transformations as 2- and 3-morphisms, respectively.

We thus deal with a homotopy action of the group O(3) on the fully dualizable objects of $\mathcal{V}$, which are in fact fusion categories: Points in O(3) give self-equivalences $\mathbb{K}[\mathcal{B}imod^{\,\mathrm{f.d.}}] \to \mathbb{K}[\mathcal{B}imod^{\,\mathrm{f.d.}}]$, paths in O(3) give natural transformations between self-equivalences, etc. The homotopy groups of the Lie group O(3) are as follows: The group of connected components is $\pi_0(\mathrm{O}(3)) = \mathbb{Z}_2$; the nontrivial component acts on monoidal categories as $(\mathcal{C}, \otimes) \longmapsto (\mathcal{C}, \otimes^{\mathrm{opp}})$. The fundamental group is $\pi_1(\mathrm{O}(3)) = \mathbb{Z}_2$; its nontrivial element acts on fusion categories as an autoequivalence, which turns out to be given by the double dual functor $(-)^{\vee\vee}$. Thus from the group-theoretical fact that the nontrivial element of $\pi_1(\mathrm{O}(3)) = \mathbb{Z}_2$ has order 2 one concludes that the quadruple dual $(-)^{\vee\vee\vee\vee}$ is trivial, which is nothing but Radford's theorem. For more details we refer to [**DSS**].

Bibliography

[Ab] Lowell Abrams, *Two-dimensional topological quantum field theories and Frobenius algebras*, J. Knot Theory Ramifications **5** (1996), no. 5, 569–587, DOI 10.1142/S0218216596000333. MR1414088

[AdHS] Jiří Adámek, Horst Herrlich, and George E. Strecker, *Abstract and concrete categories: The joy of cats*, Pure and Applied Mathematics (New York), John Wiley & Sons, Inc., New York, 1990. A Wiley-Interscience Publication. MR1051419

[ArCRW] Eddy Ardonne, Meng Cheng, Eric C. Rowell, and Zhenghan Wang, *Classification of metaplectic modular categories*, J. Algebra **466** (2016), 141–146, DOI 10.1016/j.jalgebra.2016.08.001. MR3541678

[AyFR] David Ayala, John Francis, and Nick Rozenblyum, *Factorization homology I: Higher categories*, Adv. Math. **333** (2018), 1042–1177, DOI 10.1016/j.aim.2018.05.031. MR3818096

[BaeD] John C. Baez and James Dolan, *Higher-dimensional algebra and topological quantum field theory*, J. Math. Phys. **36** (1995), no. 11, 6073–6105, DOI 10.1063/1.531236. MR1355899

[BaeHW] John C. Baez, Alexander E. Hoffnung, and Christopher D. Walker, *Higher dimensional algebra VII: Groupoidification*, Theory Appl. Categ. **24** (2010), No. 18, 489–553. MR2770073

[BaeS] J. Baez and M. Stay, *Physics, topology, logic and computation: a Rosetta Stone*, New structures for physics, Lecture Notes in Phys., vol. 813, Springer, Heidelberg, 2011, pp. 95–172, DOI 10.1007/978-3-642-12821-9_2. MR2767046

[BakK] Bojko Bakalov and Alexander Kirillov Jr., *Lectures on tensor categories and modular functors*, University Lecture Series, vol. 21, American Mathematical Society, Providence, RI, 2001, DOI 10.1090/ulect/021. MR1797619

[BarFG] John W. Barrett, João Faria Martins, and J. Manuel García-Islas, *Observables in the Turaev-Viro and Crane-Yetter models*, J. Math. Phys. **48** (2007), no. 9, 093508, 18, DOI 10.1063/1.2759440. MR2355097

[BarW] John W. Barrett and Bruce W. Westbury, *Invariants of piecewise-linear 3-manifolds*, Trans. Amer. Math. Soc. **348** (1996), no. 10, 3997–4022, DOI 10.1090/S0002-9947-96-01660-1. MR1357878

[Bar] B.H. Bartlett, *Three-dimensional TQFTs via string-nets and two-dimensional surgery*, Preprint, `arXiv:2206.13262 [math.QA]`.

[BDSV] B. Bartlett, C.L. Douglas, C.J. Schommer-Pries, and J. Vicary, *Modular categories as representations of the 3-dimensional bordism category*, Preprint, `arXiv:1509.06811 [math.AT]`.

[BeBJ] David Ben-Zvi, Adrien Brochier, and David Jordan, *Integrating quantum groups over surfaces*, J. Topol. **11** (2018), no. 4, 874–917, DOI 10.1112/topo.12072. MR3847209

[BhBF+] Lakshya Bhardwaj, Lea E. Bottini, Ludovic Fraser-Taliente, Liam Gladden, Dewi S. W. Gould, Arthur Platschorre, and Hannah Tillim, *Lectures on generalized symmetries*, Phys. Rep. **1051** (2024), 1–87, DOI 10.1016/j.physrep.2023.11.002. MR4672198

[BCGP] Christian Blanchet, Francesco Costantino, Nathan Geer, and Bertrand Patureau-Mirand, *Non-semi-simple TQFTs, Reidemeister torsion and Kashaev's invariants*, Adv. Math. **301** (2016), 1–78, DOI 10.1016/j.aim.2016.06.003. MR3539369

[BlW] Wade Bloomquist and Zhenghan Wang, *On topological quantum computing with mapping class group representations*, J. Phys. A **52** (2019), no. 1, 015301, 23, DOI 10.1088/1751-8121/aaeea1. MR3899740

[BrC] Jacob C. Bridgeman and Christopher T. Chubb, *Hand-waving and interpretive dance: an introductory course on tensor networks*, J. Phys. A **50** (2017), no. 22, 223001, 61, DOI 10.1088/1751-8121/aa6dc3. MR3659106

[Bro] Ronald Brown, *Topology and groupoids*, BookSurge, LLC, Charleston, SC, 2006. Third edition of *Elements of modern topology* [McGraw-Hill, New York, 1968; MR0227979]; With 1 CD-ROM (Windows, Macintosh and UNIX). MR2273730

[BuMCA] Oliver Buerschaper, Juan Martín Mombelli, Matthias Christandl, and Miguel Aguado, *A hierarchy of topological tensor network states*, J. Math. Phys. **54** (2013), no. 1, 012201, 46, DOI 10.1063/1.4773316. MR3059877

[ChTSR] C.-K. Chiu, J.C.Y. Teo, A.P. Schnyder, and S. Ryu, *Classification of topological quantum matter with symmetries*, Rev. Mod. Phys. 88 (2016) 035005_1–63.

[Ca] Nils Carqueville, *Orbifolds of topological quantum field theories*, Encyclopedia of mathematical physics. Vol. 3. Integrable systems & algebra, Academic Press, Amsterdam, [2025] ©2025, pp. 618–634. MR4975784

[CiPSV] J. Ignacio Cirac, David Pérez-García, Norbert Schuch, and Frank Verstraete, *Matrix product states and projected entangled pair states: concepts, symmetries, theorems*, Rev. Modern Phys. **93** (2021), no. 4, 045003, 65, DOI 10.1103/revmodphys.93.045003. MR4369910

[CGP] Francesco Costantino, Nathan Geer, and Bertrand Patureau-Mirand, *Quantum invariants of 3-manifolds via link surgery presentations and non-semi-simple categories*, J. Topol. **7** (2014), no. 4, 1005–1053, DOI 10.1112/jtopol/jtu006. MR3286896

[Co] Kevin Costello, *Topological conformal field theories and Calabi-Yau categories*, Adv. Math. **210** (2007), no. 1, 165–214, DOI 10.1016/j.aim.2006.06.004. MR2298823

[DaMNO] Alexei Davydov, Michael Müger, Dmitri Nikshych, and Victor Ostrik, *The Witt group of non-degenerate braided fusion categories*, J. Reine Angew. Math. **677** (2013), 135–177, DOI 10.1515/crelle.2012.014. MR3039775

[DaNR] Sorin Dăscălescu, Constantin Năstăsescu, and Şerban Raianu, *Hopf algebras: an introduction*, Monographs and Textbooks in Pure and Applied Mathematics, vol. 235, Marcel Dekker, Inc., New York, 2001. MR1786197

[Dec] Thibault D. Décoppet, *Drinfeld centers and Morita equivalence classes of fusion 2-categories*, Compos. Math. **161** (2025), no. 2, 305–340, DOI 10.1112/S0010437X24007553. MR4920627

[Del] P. Deligne, *Catégories tannakiennes* (French), The Grothendieck Festschrift, Vol. II, Progr. Math., vol. 87, Birkhäuser Boston, Boston, MA, 1990, pp. 111–195. MR1106898

[DGGPR] Marco De Renzi, Azat M. Gainutdinov, Nathan Geer, Bertrand Patureau-Mirand, and Ingo Runkel, *3-Dimensional TQFTs from non-semisimple modular categories*, Selecta Math. (N.S.) **28** (2022), no. 2, Paper No. 42, 60, DOI 10.1007/s00029-021-00737-z. MR4370375

[DiW] Robbert Dijkgraaf and Edward Witten, *Topological gauge theories and group cohomology*, Comm. Math. Phys. **129** (1990), no. 2, 393–429. MR1048699

[DoR] C.L. Douglas and D.J. Reutter, *Fusion 2-categories and a state-sum invariant for 4-manifolds*, Preprint, `arXiv:1812.11933 [math.QA]`.

[DSS] Christopher L. Douglas, Christopher Schommer-Pries, and Noah Snyder, *Dualizable tensor categories*, Mem. Amer. Math. Soc. **268** (2020), no. 1308, vii+88, DOI 10.1090/memo/1308. MR4254952

[DuF] David S. Dummit and Richard M. Foote, *Abstract algebra*, 3rd ed., John Wiley & Sons, Inc., Hoboken, NJ, 2004. MR2286236

[EiM] Samuel Eilenberg and Saunders MacLane, *Cohomology theory of Abelian groups and homotopy theory. I*, Proc. Nat. Acad. Sci. U.S.A. **36** (1950), 443–447, DOI 10.1073/pnas.36.8.443. MR38072

[EtGNO] Pavel Etingof, Shlomo Gelaki, Dmitri Nikshych, and Victor Ostrik, *Tensor categories*, Mathematical Surveys and Monographs, vol. 205, American Mathematical Society, Providence, RI, 2015, DOI 10.1090/surv/205. MR3242743

[EtNOM] Pavel Etingof, Dmitri Nikshych, and Victor Ostrik, *Fusion categories and homotopy theory: With an appendix by Ehud Meir*, Quantum Topol. **1** (2010), no. 3, 209–273, DOI 10.4171/QT/6. MR2677836

[FadY] L. D. Faddeev and O. A. Yakubovskiĭ, *Lectures on quantum mechanics for mathematics students: Translated from the 1980 Russian original by Harold McFaden; With an appendix by Leon Takhtajan*, Student Mathematical Library, vol. 47, American Mathematical Society, Providence, RI, 2009, DOI 10.1090/stml/047. MR2492178

[FaM] Benson Farb and Dan Margalit, *A primer on mapping class groups*, Princeton Mathematical Series, vol. 49, Princeton University Press, Princeton, NJ, 2012. MR2850125

[FaS] Julian Farnsteiner and Christoph Schweigert, *Frobenius-Schur indicators and the mapping class group of the torus*, Lett. Math. Phys. **112** (2022), no. 2, Paper No. 39, 30, DOI 10.1007/s11005-022-01513-6. MR4406833

[FeFFS] Giovanni Felder, Jürg Fröhlich, Jürgen Fuchs, and Christoph Schweigert, *Correlation functions and boundary conditions in rational conformal field theory and three-dimensional topology*, Compositio Math. **131** (2002), no. 2, 189–237, DOI 10.1023/A:1014903315415. MR1898435

[Fr1] Daniel S. Freed, *Classical Chern-Simons theory. I*, Adv. Math. **113** (1995), no. 2, 237–303, DOI 10.1006/aima.1995.1039. MR1337109

[Fr2] Daniel S. Freed, *The cobordism hypothesis*, Bull. Amer. Math. Soc. (N.S.) **50** (2013), no. 1, 57–92, DOI 10.1090/S0273-0979-2012-01393-9. MR2994995

[FrHLT] Daniel S. Freed, Michael J. Hopkins, Jacob Lurie, and Constantin Teleman, *Topological quantum field theories from compact Lie groups*, A celebration of the mathematical legacy of Raoul Bott, CRM Proc. Lecture Notes, vol. 50, Amer. Math. Soc., Providence, RI, 2010, pp. 367–403, DOI 10.1090/crmp/050/26. MR2648901

[FrQ] Daniel S. Freed and Frank Quinn, *Chern-Simons theory with finite gauge group*, Comm. Math. Phys. **156** (1993), no. 3, 435–472. MR1240583

[FrT] Daniel S. Freed and Constantin Teleman, *Gapped boundary theories in three dimensions*, Comm. Math. Phys. **388** (2021), no. 2, 845–892, DOI 10.1007/s00220-021-04192-x. MR4334249

[FreB] Edward Frenkel and David Ben-Zvi, *Vertex algebras and algebraic curves*, 2nd ed., Mathematical Surveys and Monographs, vol. 88, American Mathematical Society, Providence, RI, 2004, DOI 10.1090/surv/088. MR2082709

[Frey] P. Freyd, *Abelian Categories*. Harper and Row 1964 (see also `http://www.emis.de/journals/TAC/reprints/articles/3`).

[FroFRS] Jürg Fröhlich, Jürgen Fuchs, Ingo Runkel, and Christoph Schweigert, *Kramers-Wannier duality from conformal defects*, Phys. Rev. Lett. **93** (2004), no. 7, 070601, 4, DOI 10.1103/PhysRevLett.93.070601. MR2115752

[FrohK] J. Fröhlich and T. Kerler, *Universality in quantum Hall systems*, Nuclear Phys. B **354** (1991), no. 2-3, 369–417, DOI 10.1016/0550-3213(91)90360-A. MR1103058

[FScS] Jürgen Fuchs, Gregor Schaumann, and Christoph Schweigert, *A modular functor from state sums for finite tensor categories and their bimodules*, Theory Appl. Categ. **38** (2022), Paper No. 15, 436–594. MR4403279

[FS1] Jürgen Fuchs and Christoph Schweigert, *Symmetries, Lie algebras and representations*, Cambridge Monographs on Mathematical Physics, Cambridge University Press, Cambridge, 1997. A graduate course for physicists. MR1473220

[FS2] Jürgen Fuchs and Christoph Schweigert, *Symmetries and defects in three-dimensional topological field theory*, String-Math 2014, Proc. Sympos. Pure Math., vol. 93, Amer. Math. Soc., Providence, RI, 2016, pp. 21–40, DOI 10.1090/pspum/093/01554. MR3524232

[FS3] Jürgen Fuchs and Christoph Schweigert, *Consistent systems of correlators in non-semisimple conformal field theory*, Adv. Math. **307** (2017), 598–639, DOI 10.1016/j.aim.2016.11.020. MR3590526

[FSS] Jürgen Fuchs, Christoph Schweigert, and Carl Stigner, *Higher genus mapping class group invariants from factorizable Hopf algebras*, Adv. Math. **250** (2014), 285–319, DOI 10.1016/j.aim.2013.09.019. MR3122169

[FSV1] Jürgen Fuchs, Christoph Schweigert, and Alessandro Valentino, *Bicategories for boundary conditions and for surface defects in 3-d TFT*, Comm. Math. Phys. **321** (2013), no. 2, 543–575, DOI 10.1007/s00220-013-1723-0. MR3063919

[FSV2] Jürgen Fuchs, Christoph Schweigert, and Alessandro Valentino, *A geometric approach to boundaries and surface defects in Dijkgraaf-Witten theories*, Comm. Math. Phys. **332** (2014), no. 3, 981–1015, DOI 10.1007/s00220-014-2067-0. MR3262619

[FSWY] Jürgen Fuchs, Christoph Schweigert, Simon Wood, and Yang Yang, *Algebraic structures in two-dimensional conformal field theory*, Encyclopedia of mathematical physics. Vol. 3. Integrable systems & algebra, Academic Press, Amsterdam, [2025] ©2025, pp. 604–617. MR4975783

[FSY] Jürgen Fuchs, Christoph Schweigert, and Yang Yang, *String-net models for pivotal bicategories*, Theory Appl. Categ. **44** (2025), Paper No. 17, 474–543. MR4916625

[FucS] Jürgen Fuchs and Carl Stigner, *On Frobenius algebras in rigid monoidal categories* (English, with English and Arabic summaries), Arab. J. Sci. Eng. Sect. C Theme Issues **33** (2008), no. 2, 175–191. MR2500035

[Fun] Louis Funar, *On mapping class groups and their TQFT representations*, Encyclopedia of mathematical physics. Vol. 3. Integrable systems & algebra, Academic Press, Amsterdam, [2025] ©2025, pp. 678–688. MR4975788

[Gar] P. Garrett, *Non-existence of tensor products of Hilbert spaces*, `http://www.math.umn.edu/~garrett/m/v/nonexistence_tensors.pdf`.

[GeePT] Nathan Geer, Bertrand Patureau-Mirand, and Vladimir Turaev, *Modified quantum dimensions and re-normalized link invariants*, Compos. Math. **145** (2009), no. 1, 196–212, DOI 10.1112/S0010437X08003795. MR2480500

[GelNN] Shlomo Gelaki, Deepak Naidu, and Dmitri Nikshych, *Centers of graded fusion categories*, Algebra Number Theory **3** (2009), no. 8, 959–990, DOI 10.2140/ant.2009.3.959. MR2587410

[Gom] P.R.S. Gomes, *An introduction to higher-form symmetries*, SciPost Physics Lecture Notes 74 (2023) 1–56.

[GomP] Robert E. Gompf and András I. Stipsicz, *4-manifolds and Kirby calculus*, Graduate Studies in Mathematics, vol. 20, American Mathematical Society, Providence, RI, 1999, DOI 10.1090/gsm/020. MR1707327

[GorPS] R. Gordon, A. J. Power, and Ross Street, *Coherence for tricategories*, Mem. Amer. Math. Soc. **117** (1995), no. 558, vi+81, DOI 10.1090/memo/0558. MR1261589

[Gur] Nick Gurski, *Coherence in three-dimensional category theory*, Cambridge Tracts in Mathematics, vol. 201, Cambridge University Press, Cambridge, 2013, DOI 10.1017/CBO9781139542333. MR3076451

[Hat] Allen Hatcher, *Algebraic topology*, Cambridge University Press, Cambridge, 2002. MR1867354

[Hau] Rune Haugseng, *Iterated spans and classical topological field theories*, Math. Z. **289** (2018), no. 3-4, 1427–1488, DOI 10.1007/s00209-017-2005-x. MR3830256

[Hei] J. Heinloth, *Notes on differentiable stacks*, Mathematisches Institut, Georg-August-Universität Göttingen: Seminars Winter Term 2004/2005, Universitätsdrucke Göttingen, Göttingen, 2005, pp. 1–32. MR2206877

[HeSV] Jan Hesse, Christoph Schweigert, and Alessandro Valentino, *Frobenius algebras and homotopy fixed points of group actions on bicategories*, Theory Appl. Categ. **32** (2017), Paper No. 18, 652–681. MR3649192

[HiM] Ralf Hinze and Dan Marsden, *Equational reasoning with lollipops, forks, cups, caps, snakes, and speedometers*, J. Log. Algebr. Methods Program. **85** (2016), no. 5, 931–951, DOI 10.1016/j.jlamp.2015.12.004. MR3545401

[Hum] Benjamin Taylor Hummon, *Surface Diagrams for Gray-Categories*, ProQuest LLC, Ann Arbor, MI, 2012. Thesis (PhD)–University of California, San Diego. MR3103738

[Hus] Dale Husemoller, *Fibre bundles*, 3rd ed., Graduate Texts in Mathematics, vol. 20, Springer-Verlag, New York, 1994, DOI 10.1007/978-1-4757-2261-1. MR1249482

[JR] Theo Johnson-Freyd and David Reutter, *Minimal nondegenerate extensions*, J. Amer. Math. Soc. **37** (2024), no. 1, 81–150, DOI 10.1090/jams/1023. MR4654609

[JS1] André Joyal and Ross Street, *The geometry of tensor calculus. I*, Adv. Math. **88** (1991), no. 1, 55–112, DOI 10.1016/0001-8708(91)90003-P. MR1113284

[JS2] André Joyal and Ross Street, *Braided tensor categories*, Adv. Math. **102** (1993), no. 1, 20–78, DOI 10.1006/aima.1993.1055. MR1250465

[Kac] Victor G. Kac, *Infinite-dimensional Lie algebras*, 3rd ed., Cambridge University Press, Cambridge, 1990, DOI 10.1017/CBO9780511626234. MR1104219

[KapS] Anton Kapustin and Natalia Saulina, *Topological boundary conditions in abelian Chern-Simons theory*, Nuclear Phys. B **845** (2011), no. 3, 393–435, DOI 10.1016/j.nuclphysb.2010.12.017. MR2755172

[Kash] Rinat Kashaev, *A course on Hopf algebras*, Universitext, Springer, Cham, [2023] ©2023, DOI 10.1007/978-3-031-26306-4. MR4592646

[Kas] Christian Kassel, *Quantum groups*, Graduate Texts in Mathematics, vol. 155, Springer-Verlag, New York, 1995, DOI 10.1007/978-1-4612-0783-2. MR1321145

[Kel] G. M. Kelly, *Doctrinal adjunction*, Category Seminar (Proc. Sem., Sydney, 1972/1973), Lecture Notes in Math., Vol. 420, Springer, Berlin-New York, 1974, pp. 257–280. MR360749

[Kir] A.A. Kirillov, *String-net model of Turaev-Viro invariants*, Preprint, `arXiv:1106.6033 [math.AT]`.

[KirO] Alexander Kirillov Jr. and Viktor Ostrik, *On a q-analogue of the McKay correspondence and the ADE classification of $\mathfrak{sl}_2$ conformal field theories*, Adv. Math. **171** (2002), no. 2, 183–227, DOI 10.1006/aima.2002.2072. MR1936496

[Kit] A. Yu. Kitaev, *Fault-tolerant quantum computation by anyons*, Ann. Physics **303** (2003), no. 1, 2–30, DOI 10.1016/S0003-4916(02)00018-0. MR1951039

[KnST] Hannes Knötzele, Christoph Schweigert, and Matthias Traube, *Twisted Drinfeld centers and framed string-nets*, Quantum Topol. **15** (2024), no. 3, 537–566, DOI 10.4171/qt/213. MR4814690

[Koc] Joachim Kock, *Frobenius algebras and 2D topological quantum field theories*, London Mathematical Society Student Texts, vol. 59, Cambridge University Press, Cambridge, 2004. MR2037238

[Koh] Toshitake Kohno, *Conformal Field Theory and Topology*, Translated from the 1998 Japanese original by the author, Transl. Math. Monogr., 210, Iwanami Ser. Mod. Math. American Mathematical Society, Providence, RI, 2002. DOI 10.1090/mmono/210. MR1905659

[LaL] J. M. F. Labastida and Carlos Lozano, *Topological quantum field theory: overview*, Encyclopedia of mathematical physics. Vol. 2. Quantum mechanics, quantization and quantum computation, Academic Press, Amsterdam, 2025, pp. 453–464. Reprinted from [MR2238867]. MR4975306

[Lan] S. Lang, *Algebra* Addison-Wesley, 1984.

[LaP] Aaron D. Lauda and Hendryk Pfeiffer, *Open-closed strings: two-dimensional extended TQFTs and Frobenius algebras*, Topology Appl. **155** (2008), no. 7, 623–666, DOI 10.1016/j.topol.2007.11.005. MR2395583

[Lei] Tom Leinster, *Higher operads, higher categories*, London Mathematical Society Lecture Note Series, vol. 298, Cambridge University Press, Cambridge, 2004, DOI 10.1017/CBO9780511525896. MR2094071

[Len] Simon D. Lentner, *Quantum groups and Nichols algebras acting on conformal field theories*, Adv. Math. **378** (2021), Paper No. 107517, 71, DOI 10.1016/j.aim.2020.107517. MR4184294

[LeW] M.A. Levin and X.G. Wen, *String-net condensation: A physical mechanism for topological phases*, Phy. Review B 71 (2005) 045110_1-21.

[Lim] Lek-Heng Lim, *Tensors in computations*, Acta Numer. **30** (2021), 555–764, DOI 10.1017/S0962492921000076. MR4298222

[LoFHSV] Laurens Lootens, Jürgen Fuchs, Jutho Haegeman, Christoph Schweigert, and Frank Verstraete, *Matrix product operator symmetries and intertwiners in string-nets with domain walls*, SciPost Phys. **10** (2021), no. 3, Paper No. 053, 42, DOI 10.21468/scipostphys.10.3.053. MR4237747

[Lur] Jacob Lurie, *On the classification of topological field theories*, Current developments in mathematics, 2008, Int. Press, Somerville, MA, 2009, pp. 129–280. MR2555928

[Lyu] V. Lyubashenko, *Ribbon abelian categories as modular categories*, J. Knot Theory Ramifications **5** (1996), no. 3, 311–403, DOI 10.1142/S0218216596000229. MR1405715

[Mac] Saunders Mac Lane, *Categories for the working mathematician*, 2nd ed., Graduate Texts in Mathematics, vol. 5, Springer-Verlag, New York, 1998. MR1712872

[Mat] Yukio Matsumoto, *An introduction to Morse theory: Translated from the 1997 Japanese original by Kiki Hudson and Masahico Saito, Iwanami Series in Modern Mathematics*, Translations of Mathematical Monographs, vol. 208, American Mathematical Society, Providence, RI, 2002, DOI 10.1090/mmono/208. MR1873233

[McG] J. McGreevy, *Generalized symmetries in condensed matter*, Ann. Rev. Cond. Mat. Phys. 14 (2023) 57–82.

[McR] Robert McRae, *Twisted modules and G-equivariantization in logarithmic conformal field theory*, Comm. Math. Phys. **383** (2021), no. 3, 1939–2019, DOI 10.1007/s00220-020-03882-2. MR4244264

[Med] A. D. Mednyh, *Determination of the number of nonequivalent coverings over a compact Riemann surface* (Russian), Dokl. Akad. Nauk SSSR **239** (1978), no. 2, 269–271. MR490616

[MiS] Michaël Mignard and Peter Schauenburg, *Modular categories are not determined by their modular data*, Lett. Math. Phys. **111** (2021), no. 3, Paper No. 60, 9, DOI 10.1007/s11005-021-01395-0. MR4254068

[MooS1] G. Moore and G. Segal, *D-branes and K-theory in 2D topological field theory*, in: *Dirichlet Branes and Mirror Symmetry*, P. Aspinwall et al. eds., p. 27–108. American Mathematical Society 2009.

[MooS2] Gregory Moore and Nathan Seiberg, *Classical and quantum conformal field theory*, Comm. Math. Phys. **123** (1989), no. 2, 177–254. MR1002038

[Mor1] Jeffrey Colin Morton, *Two-vector spaces and groupoids*, Appl. Categ. Structures **19** (2011), no. 4, 659–707, DOI 10.1007/s10485-010-9225-0. MR2818677

[Mor2] Jeffrey C. Morton, *Double bicategories and double cospans*, J. Homotopy Relat. Struct. **4** (2009), no. 1, 389–428. MR2591970

[Mor3] Jeffrey C. Morton, *Cohomological twisting of 2-linearization and extended TQFT*, J. Homotopy Relat. Struct. **10** (2015), no. 2, 127–187, DOI 10.1007/s40062-013-0047-2. MR3348259

[MüW] Lukas Müller and Lukas Woike, *Dimensional reduction, extended topological field theories and orbifoldization*, Bull. Lond. Math. Soc. **53** (2021), no. 2, 392–403, DOI 10.1112/blms.12427. MR4239182

[NgS1] Siu-Hung Ng and Peter Schauenburg, *Higher Frobenius-Schur indicators for pivotal categories*, Hopf algebras and generalizations, Contemp. Math., vol. 441, Amer. Math. Soc., Providence, RI, 2007, pp. 63–90, DOI 10.1090/conm/441/08500. MR2381536

[NgS2] Siu-Hung Ng and Peter Schauenburg, *Congruence subgroups and generalized Frobenius-Schur indicators*, Comm. Math. Phys. **300** (2010), no. 1, 1–46, DOI 10.1007/s00220-010-1096-6. MR2725181

[Oru] Román Orús, *A practical introduction to tensor networks: matrix product states and projected entangled pair states*, Ann. Physics **349** (2014), 117–158, DOI 10.1016/j.aop.2014.06.013. MR3244614

[Oga] Yoshiko Ogata, *Classification of gapped ground state phases in quantum spin systems*, ICM—International Congress of Mathematicians. Vol. 5. Sections 9–11, EMS Press, Berlin, [2023] ©2023, pp. 4142–4161. MR4680397

[Ost] Viktor Ostrik, *Module categories over the Drinfeld double of a finite group*, Int. Math. Res. Not. **27** (2003), 1507–1520, DOI 10.1155/S1073792803205079. MR1976233

[PR] G. Ponzano and T. Regge, *Semiclassical limit of Racah coefficients*, In: *Spectroscopic and Group Theoretical Methods in Physics* (F. Bloch et al., eds.). North Holland Publishing Company, 1968, p. 1–58.

[Pst] Piotr Pstrągowski, *On dualizable objects in monoidal bicategories*, Theory Appl. Categ. **38** (2022), Paper No. 9, 257–310. MR4373026

[ReT] N. Reshetikhin and V. G. Turaev, *Invariants of 3-manifolds via link polynomials and quantum groups*, Invent. Math. **103** (1991), no. 3, 547–597, DOI 10.1007/BF01239527. MR1091619

[Ri1] Emily Riehl, *Categorical homotopy theory*, New Mathematical Monographs, vol. 24, Cambridge University Press, Cambridge, 2014, DOI 10.1017/CBO9781107261457. MR3221774

[Ri2] Emily Riehl, *Category theory in context*, Aurora Dover Modern Math Originals, Dover Publications, Inc., Mineola, NY, 2016. MR4727501

[Ru] Ingo Runkel, *A braided monoidal category for free super-bosons*, J. Math. Phys. **55** (2014), no. 4, 041702, 59, DOI 10.1063/1.4868467. MR3390573

[RuFFS] Ingo Runkel, Jens Fjelstad, Jürgen Fuchs, and Christoph Schweigert, *Topological and conformal field theory as Frobenius algebras*, Categories in algebra, geometry and mathematical physics, Contemp. Math., vol. 431, Amer. Math. Soc., Providence, RI, 2007, pp. 225–247, DOI 10.1090/conm/431/08275. MR2342831

[Saf] P. Safronov, *Topological quantum field theories,* Lecture notes. (see `http://sites.google.com/site/psafronov/teaching`).

[Sak] Takashi Sakai, *Riemannian geometry: Translated from the 1992 Japanese original by the author*, Translations of Mathematical Monographs, vol. 149, American Mathematical Society, Providence, RI, 1996, DOI 10.1090/mmono/149. MR1390760

[SaW] Paolo Salvatore and Nathalie Wahl, *Framed discs operads and Batalin-Vilkovisky algebras*, Q. J. Math. **54** (2003), no. 2, 213–231, DOI 10.1093/qjmath/54.2.213. MR1989873

[Scha] Sakura Schäfer-Nameki, *ICTP lectures on (non-)invertible generalized symmetries*, Phys. Rep. **1063** (2024), 1–55, DOI 10.1016/j.physrep.2024.01.007. MR4703407

[Sche] C. Scheimbauer, *Factorization homology as a fully extended topological field theory*, PhD thesis, Zürich 2014.

[Scho] Christopher John Schommer-Pries, *The classification of two-dimensional extended topological field theories*, ProQuest LLC, Ann Arbor, MI, 2009. Thesis (PhD)–University of California, Berkeley. MR2713992

[ScW] Christoph Schweigert and Lukas Woike, *Orbifold construction for topological field theories*, J. Pure Appl. Algebra **223** (2019), no. 3, 1167–1192, DOI 10.1016/j.jpaa.2018.05.020. MR3862670

[Seg] Graeme Segal, *The definition of conformal field theory*, Topology, geometry and quantum field theory, London Math. Soc. Lecture Note Ser., vol. 308, Cambridge Univ. Press, Cambridge, 2004, pp. 421–577. MR2079383

[Sel] P. Selinger, *A survey of graphical languages for monoidal categories*, New structures for physics, Lecture Notes in Phys., vol. 813, Springer, Heidelberg, 2011, pp. 289–355, DOI 10.1007/978-3-642-12821-9_4. MR2767048

[Shi] Kenichi Shimizu, *Non-degeneracy conditions for braided finite tensor categories*, Adv. Math. **355** (2019), 106778, 36, DOI 10.1016/j.aim.2019.106778. MR3996323

[SkY1] Andrzej Skowroński and Kunio Yamagata, *A general form of non-Frobenius self-injective algebras*, Colloq. Math. **105** (2006), no. 1, 135–141, DOI 10.4064/cm105-1-12. MR2242504

[SkY2] Andrzej Skowroński and Kunio Yamagata, *Frobenius algebras. I: Basic representation theory*, EMS Textbooks in Mathematics, European Mathematical Society (EMS), Zürich, 2011, DOI 10.4171/102. MR2894798

[Sny] Noah Snyder, *Mednykh's formula via lattice topological quantum field theories*, Proceedings of the 2014 Maui and 2015 Qinhuangdao conferences in honour of Vaughan F. R. Jones' 60th birthday, Proc. Centre Math. Appl. Austral. Nat. Univ., vol. 46, Austral. Nat. Univ., Canberra, 2017, pp. 389–398. MR3635678

[SöV] Kürşat Sözer and Alexis Virelizier, *3d TQFTs and 3-manifold invariants*, Encyclopedia of mathematical physics. Vol. 3. Integrable systems & algebra, Academic Press, Amsterdam, [2025] ©2025, pp. 529–541. MR4975777

[Swe] Moss E. Sweedler, *Hopf algebras*, Mathematics Lecture Note Series, W. A. Benjamin, Inc., New York, 1969. MR252485

[Tak] Mitsuhiro Takeuchi, *Free Hopf algebras generated by coalgebras*, J. Math. Soc. Japan **23** (1971), 561–582, DOI 10.2969/jmsj/02340561. MR292876

[Til] Ulrike Tillmann, *$\mathcal{S}$-structures for k-linear categories and the definition of a modular functor*, J. London Math. Soc. (2) **58** (1998), no. 1, 208–228, DOI 10.1112/S0024610798006383. MR1670122

[Tur1] V. G. Turaev, *Quantum invariants of knots and 3-manifolds*, De Gruyter Studies in Mathematics, vol. 18, Walter de Gruyter & Co., Berlin, 1994. MR1292673

[Tur2] Vladimir Turaev, *Homotopy quantum field theory*, EMS Tracts in Mathematics, vol. 10, European Mathematical Society (EMS), Zürich, 2010. Appendix 5 by Michael Müger and Appendices 6 and 7 by Alexis Virelizier, DOI 10.4171/086. MR2674592

[TV1] Vladimir Turaev and Alexis Virelizier, *Monoidal categories and topological field theory*, Progress in Mathematics, vol. 322, Birkhäuser/Springer, Cham, 2017, DOI 10.1007/978-3-319-49834-8. MR3674995

[TV2] V. G. Turaev and O. Ya. Viro, *State sum invariants of 3-manifolds and quantum 6j-symbols*, Topology **31** (1992), no. 4, 865–902, DOI 10.1016/0040-9383(92)90015-A. MR1191386

[Wah] N. Wahl, *Ribbon braids and related operads*, PhD thesis, 2001.

[Wal] Konrad Waldorf, *More morphisms between bundle gerbes*, Theory Appl. Categ. **18** (2007), No. 9, 240–273. MR2318389

[Wei] Charles A. Weibel, *An introduction to homological algebra*, Cambridge Studies in Advanced Mathematics, vol. 38, Cambridge University Press, Cambridge, 1994, DOI 10.1017/CBO9781139644136. MR1269324

[Wen] Xiao-Gang Wen, *Colloquium: zoo of quantum-topological phases of matter*, Rev. Modern Phys. **89** (2017), no. 4, 041004, 17, DOI 10.1103/RevModPhys.89.041004. MR3746348

[Wil] Simon Willerton, *The twisted Drinfeld double of a finite group via gerbes and finite groupoids*, Algebr. Geom. Topol. **8** (2008), no. 3, 1419–1457, DOI 10.2140/agt.2008.8.1419. MR2443249

[Wit] Edward Witten, *Quantum field theory and the Jones polynomial*, Comm. Math. Phys. **121** (1989), no. 3, 351–399. MR990772

Index

80 **Jürgen Fuchs, Christoph Schweigert, and Lukas Woike,** A First Course in Topological Field Theory, 2026

79 **Adam Bowers,** Nigel Kalton's Lectures in Nonlinear Functional Analysis, 2024

78 **Paul B. Larson,** Extensions of the Axiom of Determinacy, 2023

77 **Imre Bárány,** Combinatorial Convexity, 2021

76 **Mario Garcia-Fernandez and Jeffrey Streets,** Generalized Ricci Flow, 2021

75 **Raymond Cheng, Javad Mashreghi, and William T. Ross,** Function Theory and ℓ^p Spaces, 2020

74 **Leonid Polterovich, Daniel Rosen, Karina Samvelyan, and Jun Zhang,** Topological Persistence in Geometry and Analysis, 2020

73 **Armand Borel,** Introduction to Arithmetic Groups, 2019

72 **Pavel Mnev,** Quantum Field Theory: Batalin–Vilkovisky Formalism and Its Applications, 2019

71 **Alexander Grigor'yan,** Introduction to Analysis on Graphs, 2018

70 **Ian F. Putnam,** Cantor Minimal Systems, 2018

69 **Corrado De Concini and Claudio Procesi,** The Invariant Theory of Matrices, 2017

68 **Antonio Auffinger, Michael Damron, and Jack Hanson,** 50 Years of First-Passage Percolation, 2017

67 **Sylvie Ruette,** Chaos on the Interval, 2017

66 **Robert Steinberg,** Lectures on Chevalley Groups, 2016

65 **Alexander M. Olevskii and Alexander Ulanovskii,** Functions with Disconnected Spectrum, 2016

64 **Larry Guth,** Polynomial Methods in Combinatorics, 2016

63 **Gonçalo Tabuada,** Noncommutative Motives, 2015

62 **H. Iwaniec,** Lectures on the Riemann Zeta Function, 2014

61 **Jacob P. Murre, Jan Nagel, and Chris A. M. Peters,** Lectures on the Theory of Pure Motives, 2013

60 **William H. Meeks III and Joaquín Pérez,** A Survey on Classical Minimal Surface Theory, 2012

59 **Sylvie Paycha,** Regularised Integrals, Sums and Traces, 2012

58 **Peter D. Lax and Lawrence Zalcman,** Complex Proofs of Real Theorems, 2012

57 **Frank Sottile,** Real Solutions to Equations from Geometry, 2011

56 **A. Ya. Helemskii,** Quantum Functional Analysis, 2010

55 **Oded Goldreich,** A Primer on Pseudorandom Generators, 2010

54 **John M. Mackay and Jeremy T. Tyson,** Conformal Dimension, 2010

53 **John W. Morgan and Frederick Tsz-Ho Fong,** Ricci Flow and Geometrization of 3-Manifolds, 2010

52 **Marian Aprodu and Jan Nagel,** Koszul Cohomology and Algebraic Geometry, 2010

51 **J. Ben Hough, Manjunath Krishnapur, Yuval Peres, and Bálint Virág,** Zeros of Gaussian Analytic Functions and Determinantal Point Processes, 2009

50 **John T. Baldwin,** Categoricity, 2009

49 **József Beck,** Inevitable Randomness in Discrete Mathematics, 2009

48 **Achill Schürmann,** Computational Geometry of Positive Definite Quadratic Forms, 2008

47 **Ernst Kunz, David A. Cox, and Alicia Dickenstein,** Residues and Duality for Projective Algebraic Varieties, 2008

46 **Lorenzo Sadun,** Topology of Tiling Spaces, 2008

45 **Matthew Baker, Brian Conrad, Samit Dasgupta, Kiran S. Kedlaya, and Jeremy Teitelbaum,** p-adic Geometry, 2008

44 **Vladimir Kanovei,** Borel Equivalence Relations, 2008

43 **Giuseppe Zampieri,** Complex Analysis and CR Geometry, 2008

Printed and bound by CPI Group (UK) Ltd, Croydon, CR0 4YY
23/08/2026
14947093-0001